I0875881

Sustainability for Engineering and Science

Sustainability for Engineering and Science provides an interdisciplinary overview of key sustainability challenges and solutions within the framework of the United Nations 2030 Agenda for Sustainable Development. Drawing on the expertise of professors from different disciplines at Sapienza University of Rome, the book explores energy and environmental sustainability, raw materials and the circular economy, sustainable mobility, hydroclimatology and water cycle sustainability, ecology and ecosystem services, and the legal framework of sustainable development.

By bringing together environmental, technological, and regulatory perspectives in a single volume, the book highlights the interconnections between natural resources, human activities, and sustainable development. Combining an accessible approach with scientific and technical rigor, it is designed primarily for university students from different academic backgrounds, while also providing a valuable resource for educators, researchers, professionals, and anyone seeking a clear and integrated understanding of sustainability challenges and solutions.

Sustainability for Engineering and Science

A Multidisciplinary Approach

Silvia Serranti, Francesco Cioffi,
Livio De Santoli, Angelo Lalli,
Fausto Manes, Luca Persia,
Davide Shingo Usami, and
Davide Astiaso Garcia

Routledge
Taylor & Francis Group
LONDON AND NEW YORK

Cover Image: © Shutterstock.com

First published 2027
by Routledge
4 Park Square, Milton Park, Abingdon, Oxon OX14 4RN

and by Routledge
605 Third Avenue, New York, NY 10158

Routledge is an imprint of the Taylor & Francis Group, an informa business

British Library Cataloguing-in-Publication Data
A catalogue record for this book is available from the British Library

ISBN: 978-1-032-97837-6 (hbk)
ISBN: 978-1-032-97574-0 (pbk)
ISBN: 978-1-003-59570-0 (ebk)

DOI: 10.1201/9781003595700

Typeset in Sabon
by SPi Technologies India Pvt Ltd (Straive)

Contents

Author Biography

Silvia Serranti is a Full Professor of Raw Materials Engineering at Sapienza University of Rome, Department of Chemical Engineering, Materials, and Environment (DICMA). She is a member of the University Technical-Scientific Committee on Sustainability and serves as Sapienza's representative in the "Resources and Waste" working group within the Italian Universities Network for Sustainable Development (RUS). She is president of the Civil and Industrial Engineering Teaching Area Council (CAD-ICI) at the Latina campus, coordinating the BSc in Environmental Engineering for Sustainable Development and the MSc in Green Industrial Engineering for Sustainable Development. Her research activity focuses on the circular economy, sustainable resource management, and urban mining, with particular expertise in the application of advanced spectroscopic and imaging techniques for the characterization and valorization of waste and secondary raw materials. She coordinates the Raw Materials Laboratory (RawMaLab) and has authored more than 200 peer-reviewed papers indexed in Scopus. She has been involved in numerous national and European research projects as both a coordinator and participant. Her work aims to foster a materials-for-sustainability approach, bridging research, education, and industrial implementation to support the transition toward a circular and resource-efficient future.

Francesco Cioffi is a former Associate Professor of Hydraulics at the Department of Civil, Building, and Environmental Engineering of Sapienza University of Rome, where he served until his retirement on November 1, 2025. His research interests include experimental hydraulics, with a particular focus on laboratory and field measurements of velocity and turbulence; environmental hydraulics, including hydraulic and water-quality modeling and solid transport processes; and applied hydrology, with an emphasis on the relationships between floods, droughts, and climate change. His current research activity focuses on changes in rainfall patterns over the last century; statistical downscaling methodologies for projecting rainfall, temperature, and sea level in coastal areas under global warming scenarios; forecasting models for tropical cyclone rainfall and flood events; and the development of early-warning systems for floods and droughts. Over the course of his career, he has collaborated on more than a dozen research projects funded by Italian public institutions. These projects addressed a wide range of topics, including the control and management of coastal erosion processes, the design and environmental assessment of inshore and offshore aquaculture systems, the recovery and management of water quality in lagoons, rivers, and lakes; the control of tidal dynamics in the Venice Lagoon; the management of solid transport in rivers and reservoirs; the development of early-warning systems for flooding; and the assessment of the climate vulnerability of crops in developing countries.

He has taught courses in fluid mechanics, environmental fluid mechanics, hydroclimatology, and sustainability science.

Livio De Santoli is a member of the Board of the Italian Authority for Energy, Environment, and Waste for the 2026–2033 term. He is a Full Professor (ING-IND/11) at Sapienza University of Rome, holding the chairs of Energy Management (Faculty of Engineering, Master of Science in Energy Engineering) and Technical Systems (Faculty of Architecture, Bachelor of Architecture). He served as Pro-Rector for Energy Policies from 2018 to 2020 and Pro-Rector for Sustainability from 2021 to 2025. From 2008 to 2017, he served as Energy Manager at Sapienza University of Rome. Other activities include Sapienza University's representative in the Rete delle Università Sostenibili (RUS); coordinator of the Network of Sustainable Universities; Dean of the Faculty of Architecture Valle Giulia; Director of the CITERA Research Centre; Director of the Master "Strategy Energy Management Systems"; Member of the Scientific Committee of the National Energy Cluster; President of the Steering Committee of the "FREE Coordination"; President of AiCARR (Italian Association of Air Conditioning, Heating, and Refrigeration); and Member of the International Advisory Board of the international journal *Building Services Engineering Research & Technology* (UK). He has been president of the ASN Commission for Macroarea C2 (National Scientific Qualification). His activities in energy strategy include the Energy Plan for Rome, the Energy Plan for the Molise Region, the Energy Plan for Sapienza University of Rome, and the master plans of Castel Del Giudice and Civitavecchia. He is the author of more than 250 papers in the fields of energy management, RES, and energy planning, as well as 15 university books and 5 essays.

Angelo Lalli is a distinguished academic and professional in administrative law, with a career spanning academia, public institutions, and scholarly activity.

He is a tenured Associate Professor at Sapienza University of Rome and a lawyer qualified to appear before the higher courts. He has also obtained the national qualification for a full professorship. He carries out extensive academic work, teaching administrative law, environmental and sustainability law, and public regulation of the economy. He is a member of Sapienza's Technical-Scientific Committee for Sustainability. His institutional experience is significant: he has worked at the Italian Competition Authority, where he held senior positions, and at the Presidency of the Council of Ministers during the Monti government, dealing with legislative and regulatory activities. He has also taken part in ministerial commissions and national working groups on environmental, economic, and bioeconomy issues. He currently serves as a consultant to the Technical Mission Structure at the Ministry of Infrastructure and Transport, where he is involved in strategic guidance, planning, analysis, research, design, and specialized advisory activities concerning the country's key infrastructures (motorways, railways, ports, airports, and social housing). He is also an advisor to the Minister of Health. He is the author of numerous academic publications, monographs, and edited volumes on administrative law, competition, the environment, and digital technologies. He regularly participates as a speaker at national and international conferences. His profile reflects a strong integration of academic research, institutional activity, and professional engagement.

Fausto Manes is a Full Professor in the Department of Environmental Biology at Sapienza University of Rome. He retired on November 1, 2021, and was appointed Professor Emeritus of Ecology by decree of the Minister of University and Research on January 19, 2023. He was the holder of the Ecology and Applied Ecology courses. He has also taught the following courses: Botany, Agricultural Chemistry, Plant Ecology, Complements of Applied Ecology, Urban and Environmental Ecology, and Ecology and Sustainable Development. His research activity is focused on the response of natural and crop Mediterranean plant species to environmental stress factors (atmospheric pollutants and climate change); ▪ Analysis of vegetation and environmental quality in urban and natural areas, and quantification of the regulating ecosystem services provided by vegetation; ▪ Development of models for ecosystem analysis, simulation, and mapping of structural features, functional processes, and ecosystem services; ▪ Enhancement of green infrastructure and, more generally, of natural capital through the design of "nature-based solutions" for the improvement of the quality of life; ▪ Urban and peri-urban forests and their role in improving air quality and regulating local climate for the well-being of citizens; ▪ Estimation, including through the analysis of satellite remote sensing data, of the removal of atmospheric pollutants and temperature mitigation by vegetation and the monetary evaluation of the relative ecosystem service. He has authored more than 200 scientific papers in international journals, book chapters, and monographs, 149 of which are indexed in Scopus. His h-index is 31.

Luca Persia is a Professor of Transport Policies, Sustainability Science, and Road Safety at the Research Centre for Transport and Logistics (CTL) of Sapienza University of Rome, where he also serves as director. He is also Chairman of CTLup, a university spin-off startup that aims to foster innovation, technology transfer, and the market uptake of research results in the fields of transport, logistics, and sustainable mobility. With a career spanning over 25 years, he has developed extensive expertise in urban transport systems, road safety, sustainable mobility, and innovative approaches to transport planning. His work integrates research and strategic applications across international initiatives supported by the European Commission, the World Bank, the OECD, and the UNECA, often in scientific coordination or leadership roles. His research focuses on road accident analysis, safety countermeasures, sustainable transport policies, cycling mobility systems, and intelligent transport systems, supported by a multidisciplinary team within the CTL. He is actively involved in major international research and policy networks, serving as president of the Forum of European Road Safety Research Institutes (FERSI) and contributing to organizations such as IRTAD, ETSC, and ECTRI in coordination and advisory roles. He holds a Master's degree in Civil and Transport Engineering and a PhD in Transport Engineering from Sapienza University of Rome.

Davide Shingo Usami is a University Researcher at the Research Centre for Transport and Logistics at Sapienza University of Rome, with over two decades of scholarly activity in road safety and sustainable mobility. He is one of the founders of CTLup, a university spin-off startup focused on the valorization and transfer of research results into innovative solutions in transport, logistics, and sustainable mobility. His research focuses on road accident analysis, safety performance assessment, and transport system planning, with a strong emphasis on quantitative and data-driven methodologies. His work contributes to applied research activities supporting evidence-based decision-making in transport and road safety. He holds an MSc in Environmental Engineering and a PhD in Transport Engineering from Sapienza University of Rome.

Davide Astiaso Garcia is a Full Professor of Energy Systems at the Department of Electrical and Energy Engineering at Sapienza University of Rome. He is the Secretary General of the Italian Wind Energy Association (ANEV). He is the Rector's Delegate for Sustainability at Sapienza University of Rome and Sapienza's representative within the Universities Network for Sustainable Development (RUS) Agreement. His areas of expertise include, among others, wind energy, other renewable energy sources, smart energy systems, and energy efficiency in buildings. Astiaso Garcia has an extensive publication record, including more than 190 peer-reviewed papers in scientific journals indexed in the Scopus database. His publications have received over 6,500 citations in Scopus, and he has an h-index of 48.

He has a research portfolio funded with over €2.5 million through European competitive research projects, and he works as an expert for the European Commission in the evaluation of Horizon Europe project proposals. In particular, he has participated in more than 15 European and national research projects in the field of renewable energies and smart energy systems (in many cases as Principal Investigator and Project Coordinator). Since 2019, he has been included in the World's Top 2% Scientists List in the Energy field, both for the whole career and for the last year's publications, according to the "Updated science-wide author databases of standardized citation indicators" by Elsevier and SciTech Strategies (Stanford University). His research activities on the energy transition of the smaller islands were selected among the 100 Italian Renewable Energy Stories in the report promoted by Symbola—Fondazione per le qualità italiane and ENEL, in collaboration with KEY—The Energy Transition Expo, dedicated to technologies developed in the world of renewables. He is currently among the top Italian scientists, a national census of the scientists and scholars with the greatest impact measured by the h-index value. Research area: Engineering—Energy.

Preface

The depletion of natural resources, climate change, environmental pollution, and biodiversity loss are among the most critical challenges affecting our planet. The United Nations 2030 Agenda for Sustainable Development addresses these issues through 17 Sustainable Development Goals (SDGs), which require coordinated actions at global, national, and local levels. Achieving these goals is a complex challenge that requires the involvement of all stakeholders, increased awareness through education, and sustained investment in research and innovation. In this context, universities play a fundamental role in promoting the transition toward sustainability. Through education, research, and knowledge transfer, they contribute to the development of the skills and solutions needed to address interconnected environmental, social, and economic challenges.

This book originates from the interdisciplinary teaching experience developed at Sapienza University of Rome within the course "Sustainability Sciences in Engineering," offered since the academic year 2020–2021, and reflects the need for an integrated and multidisciplinary approach to sustainability. This volume is structured into six chapters, each addressing a key area related to sustainability: energy and environmental sustainability, raw materials and recycling, sustainable mobility, hydroclimatology, ecology and ecosystem services, and the legal framework of sustainable development. These topics are closely linked to the objectives of the UN 2030 Agenda and highlight the interconnections between natural systems, technological solutions, and regulatory frameworks. The authors, all professors at Sapienza University of Rome, come from different departments and disciplinary backgrounds, reflecting the interdisciplinary nature of sustainability sciences.

This book is intended primarily as an introductory resource for university students, but it may also be of interest to educators, professionals, and a broader audience seeking a clear and structured overview of sustainability challenges and solutions.

Introduction to the concept of sustainability and the UN 2030 Agenda for Sustainable Development

The concept of sustainability and sustainable development has evolved over several decades through a series of key international milestones. One of the earliest and most significant steps was the United Nations Conference on the Human Environment, held in Stockholm in 1972, which established fundamental principles concerning human rights and responsibilities toward the environment (UN, 1972).

A major advancement was achieved in 1987 with the publication of the Brundtland Report "Our Common Future" (UN, 1987), which introduced the widely accepted definition of sustainable development as "development that meets the needs of the present without compromising the ability of future generations to meet their own needs." This definition remains a cornerstone in sustainability studies.

The concept was further strengthened during the Earth Summit held in Rio de Janeiro in 1992, where global leaders adopted key documents such as the "Rio Declaration" and "Agenda 21" (UN, 1992), outlining strategies for sustainable development. In 2002, the World Summit on Sustainable Development in Johannesburg reaffirmed the importance of integrating the three fundamental pillars of sustainability: environmental protection, economic growth, and social inclusion.

A major step forward was taken on 25 September 2015, when the United Nations General Assembly adopted the "2030 Agenda for Sustainable Development" (UN, 2015), an ambitious action plan for people, planet, and prosperity, endorsed by all 193 UN Member States. This global framework includes 17 *Sustainable Development Goals* (SDGs) and 169 targets addressing the most pressing challenges facing humanity. The SDGs build upon the *Millennium Development Goals* (MDG), adopted by the United Nations General Assembly in 2000, and aim to complete what these were unable to achieve through actions in areas of crucial importance for humanity and the planet.

The 2030 Agenda is structured around five key dimensions, commonly referred to as the "5 Ps": People, Planet, Prosperity, Peace, and Partnership. These dimensions reflect a comprehensive vision of sustainable development, aiming to eradicate poverty and hunger, protect natural resources,

DOI: 10.1201/9781003595700-1

ensure inclusive and prosperous societies, promote peace and justice, and strengthen global cooperation.

The 17 SDGs are as follows:

- **SDG 1—No Poverty:** End poverty in all its forms everywhere.
- **SDG 2—Zero Hunger:** End hunger, achieve food security and improved nutrition, and promote sustainable agriculture.
- **SDG 3—Good Health and Well-being:** Ensure healthy lives and promote well-being for all at all ages.
- **SDG 4—Quality Education:** Ensure inclusive and equitable quality education and promote lifelong learning opportunities for all.
- **SDG 5—Gender Equality:** Achieve gender equality and empower all women and girls.
- **SDG 6—Clean Water and Sanitation:** Ensure availability and sustainable management of water and sanitation for all.
- **SDG 7—Affordable and Clean Energy:** Ensure access to affordable, reliable, sustainable, and modern energy for all.
- **SDG 8—Decent Work and Economic Growth:** Promote sustained, inclusive, and sustainable economic growth, full and productive employment, and decent work for all.
- **SDG 9—Industry, Innovation, and Infrastructure:** Build resilient infrastructure, promote inclusive and sustainable industrialization, and foster innovation.
- **SDG 10—Reduced Inequalities:** Reduce inequality within and among countries.
- **SDG 11—Sustainable Cities and Communities:** Make cities and human settlements inclusive, safe, resilient, and sustainable.
- **SDG 12—Responsible Consumption and Production:** Ensure sustainable consumption and production patterns.
- **SDG 13—Climate Action:** Take urgent action to combat climate change and its impacts.
- **SDG 14—Life Below Water:** Conserve and sustainably use the oceans, seas, and marine resources for sustainable development.
- **SDG 15—Life on Land:** Protect, restore, and promote sustainable use of terrestrial ecosystems; sustainably manage forests; combat desertification; halt and reverse land degradation; and halt biodiversity loss.
- **SDG 16—Peace, Justice, and Strong Institutions:** Promote peaceful and inclusive societies for sustainable development, provide access to justice for all, and build effective, accountable, and inclusive institutions at all levels.
- **SDG 17—Partnerships for the Goals:** Strengthen the means of implementation and revitalize the global partnership for sustainable development.

The 17 SDGs address a comprehensive and interconnected set of global challenges, including poverty, hunger, health, education, gender equality, water and sanitation, affordable and clean energy, decent work and economic growth, industry, innovation and infrastructure, reduced inequalities, sustainable cities and communities, responsible consumption and production, climate action, life below water, life on land, peace, justice and strong institutions, and partnerships for the goals. Each goal is associated with specific targets and indicators designed to monitor progress and guide policy actions.

Achieving these goals requires the active involvement of all stakeholders, including governments, the private sector, academia, civil society, and individual citizens. Only through a shared and coordinated effort will it be possible to ensure a sustainable, equitable, and resilient future for present and future generations.

REFERENCES

United Nations (UN) (1972) *Report of the United Nations Conference on the Human Environment*. Stockholm, 5–16 June, 1972.

United Nations (UN) (1987) *Report of the World Commission on Environment and Development: Our Common Future*. Oxford University Press, Oxford.

United Nations (UN) (1992) *Agenda 21*. United Nations, New York.

United Nations (UN) (2015) *Transforming Our World: The 2030 Agenda for Sustainable Development*. Resolution adopted by the General Assembly on 25 September, 2015.

Chapter 1

Energy and environmental sustainability: Relevant SDGs: 7, 11, 12, and 13

Livio De Santoli and Davide Astiaso Garcia

1.1 THE CHARACTERISTICS OF THE ENERGY TRANSITION

1.1.1 Decarbonizing the economy

For many years, geologists, scientists, and climatologists have debated the exact date when the Holocene, which began 11,000 years ago, ended and whether it is possible to define a different geological epoch, since for the first time in human history, the Earth's environment, in all its physical, chemical, and biological characteristics, is strongly influenced by the *effects of human action*, particularly the increase in atmospheric concentrations of CO_2 and CH_4. Although the term is not universally accepted by the scientific community, the *Anthropocene* is used to emphasize humanity's responsibility for the increasingly evident changes to ecosystems (Crutzen, 2005).

The average atmospheric concentration of CO_2 is constantly increasing over the years, as is systematically recorded by the international scientific community (Friedlingstein et al., 2020). The trend demonstrates that this phenomenon shows no signs of changing toward a downward trend: global CO_2 concentration increased from approximately 277 ppm in 1750 to 407 ppm in 2018 (a 46% increase). In particular, 2016 was the first year in which a concentration above 400 ppm was recorded.

This is due to CO_2 emissions into the atmosphere, which, over the years, have always been constantly growing, albeit at a slower rate, around an average value of 35 Gt CO_2 (billion tons of CO_2) (*International Energy Agency, IEA*), although it has been estimated that they should decrease by a value between 1 and 2 billion tons per year in the period 2020–2030 to combat climate change and comply with the Paris Agreement.

This growth is due almost entirely (90%) to the energy sector, due to pollution resulting from the use of fossil fuels (gas, oil, and coal, with emissions of approximately 7, 11, and 14 billion tons of CO_2 per year, respectively).

The energy sector's significant responsibility for climate change, however, must be placed within a comprehensive framework when addressing strategies to combat it. The overarching goal of sustainable development is the

 DOI: 10.1201/9781003595700-2

long-term stability of the economy and the environment, which can only be achieved through the integration and recognition of all environmental, economic, and social aspects in decision-making processes. This is the theme of the United Nations 2030 Agenda, which has defined a series of objectives (SDGs) explicitly aimed at protecting the planet from a holistic perspective to ensure prosperity and peace for its peoples.

Among the SDGs, goal number 7 refers to universal access to energy and its security (*clean energy for all*). Access to energy means ensuring widespread availability for all humankind, at least for basic needs, and energy security, according to the IEA definition, refers to the need for an adequate supply of energy at a reasonable price. Both conditions are linked to the concept of *democracy*, because this access and security belong to the freedom of individuals and society. *Clean energy*, then, means using non-polluting and renewable sources.

The UN's objectives extend the themes of the Universal Declaration of Human Rights to the environment and directly to energy. In Italy, too, the constitutional relevance of the value of the *environment* and *the right to the environment* has been repeatedly affirmed in various rulings by both the Constitutional Court and the Court of Cassation through an extensive interpretation of the concepts of landscape protection (Article 9) and health protection (Article 32). However, the challenging objective of sustainability, which presupposes combining protection and conservation policies with economic development policies, suggests a more explicit reference to environmental protection among the fundamental principles of the Constitution, whose correlation with energy is indirect, albeit evident. Energy should therefore acquire, in a broader and more inclusive context of different aspects (economic, social, and environmental), a system of protection on a par with other constitutional rights, which would allow it to fully qualify as a social right.

In Europe, in particular, the operational interpretation of the European Commission's *Clean Energy for All the European* regulatory package (2019) requires member states to achieve the objectives of complete decarbonization by 2050, which also includes the economic sector and not only the energy one; in fact, the decarbonization of the economy also includes social aspects, such as the elimination of all inequalities.

The two main actions, which are central to the energy transition and which must be placed at the heart of any decarbonization strategy, concern energy efficiency and the use of renewable energy sources and must be included in an innovative energy model that is completely different from the existing one, which refers to distributed energy generation and energy communities.

Energy efficiency, when applied to an energy transformation process, is the ratio between the energy output (available energy, used as a useful effect of the process) and the energy input to the process itself and coincides with the first-law efficiency of the transformation in question. The purpose of energy efficiency interventions is to impose measures on the transformation that minimize energy degradation, reducing losses due to the irreversibility of its transformations.

A process that produces E_u (useful energy) and requires input energy E_i has an efficiency equal to:

$$\varepsilon = E_u / E_{i.}$$

If the energy efficiency of the process increases from ε_1 to ε_2, we obtain (for the same amount of useful energy produced) an energy saving equal to:

$$R = E_{i1} - E_{i2} = E_u\left[\left(1/\varepsilon_1\right) - \left(1/\varepsilon_2\right)\right].$$

Energy savings (measured in energy units) are therefore linked to the energy efficiency of the transformation process (a dimensionless value determined by a ratio of homogeneous quantities). Energy savings, of course, can also be determined by a reduction in the E_u value $(E_{u2} < E_{u1})$.

Energy efficiency and renewable energy sources are essential to achieving decarbonization goals, but only within a profoundly modified energy model compared to the current one, based on the advent of renewables, which will replace the system based on the use of fossil fuels. Distributed energy generation (typical of renewable sources, which by their nature are distributed across the territory) requires a focus on the role of the individual and their direct involvement as an energy producer and not just a consumer (in fact, we speak of *prosumers*): this is one of the reasons for the advent and importance of energy communities.

The role of the transition in the energy sector will therefore be decisive in the coming years. This transition will be a true revolution, as it will bring about a radical change in the structure of energy generation, distribution, and use and will involve a progressive adaptation of society toward the process of access to clean energy, in accordance with SDG 7, with direct reference to the process of energy democratization (De Santoli, 2021).

In the new model, existing infrastructure must become more flexible and intelligent: decentralized renewable energy sources and other trends, such as the growing number of electric vehicles and the increased electrification of energy consumption, are changing the grid load profile. The energy transition is therefore strictly dependent on a consistent improvement of its infrastructure and the transition to a global implementation of smart grids. Energy storage also plays a key role in this transformation process and therefore plays a significant role, taking on the characteristics of a true generation system, like any other energy source.

The introduction and implementation of a new energy policy by the European Commission has been very incisive over the last twenty years and has assigned Europe a significant role in the process of decarbonization of the economy, which Europe itself has decided to complete by 2050. This policy has been expressed through a series of Directives and Regulations concerning energy efficiency (EED, *Energy Efficiency Directive*), renewable sources (RED, *Renewable Energy Directive*), and energy efficiency in buildings (EPBD, *Energy Performance of Building Directive*).

European Directives and Regulations refer to strategic documents, such as *Clean Energy for All the European* and the *Green New Deal*, which are a set of policy initiatives proposed by the European Commission with the overall objective of achieving climate neutrality in Europe by 2050 and approved by the Parliament in 2020.[1] The EU Regulation, which sets binding emission reduction levels for each Member State by 2030, has defined the reduction level at 55% and, consequently, a share of energy that must come from renewable sources equal to 40% and a reduction in primary energy consumption of at least 36–39% compared to the 2007 scenario, in relation to final and primary consumption.

1.1.2 The use of renewable energy sources

The share of renewable energy has progressively increased in recent years and will be even more significant in the near future. Decarbonization in the energy sector, in fact, refers to the increase in the installed capacity of renewable energy sources and the parallel deactivation of fossil fuel energy generation capacities: in particular, the share of renewables in the energy mix must increase by 2050 from the current 30% (2016) to 80% (Figure 1.1).

In parallel with their diffusion on the market, the cost of renewable energy plants is progressively decreasing: in terms of *Levelised Cost of Energy* (LCOE) of the different energy generation technologies, which is defined as the net present value of the unit cost of energy produced by a given technology during its useful life, the decrease referred to renewable generation compared to traditional fossil generation is evident; Figure 1.2 shows this new market situation, defined as *grid parity*.

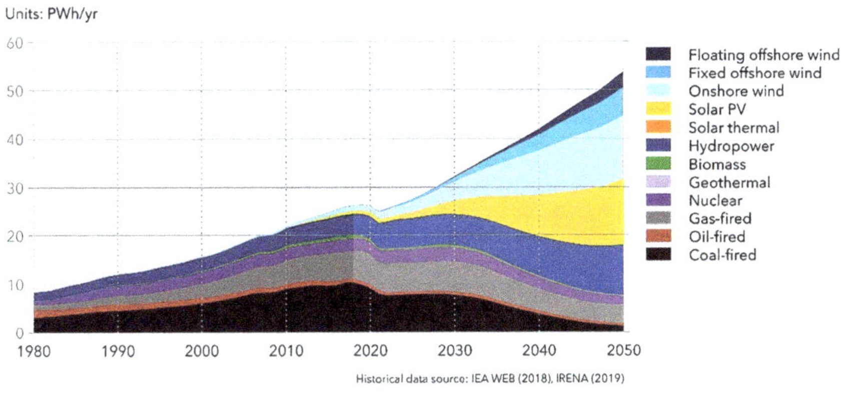

Figure 1.1 Share of energy from renewable sources in the European energy mix. **Energy Transition Outlook, 2020.**

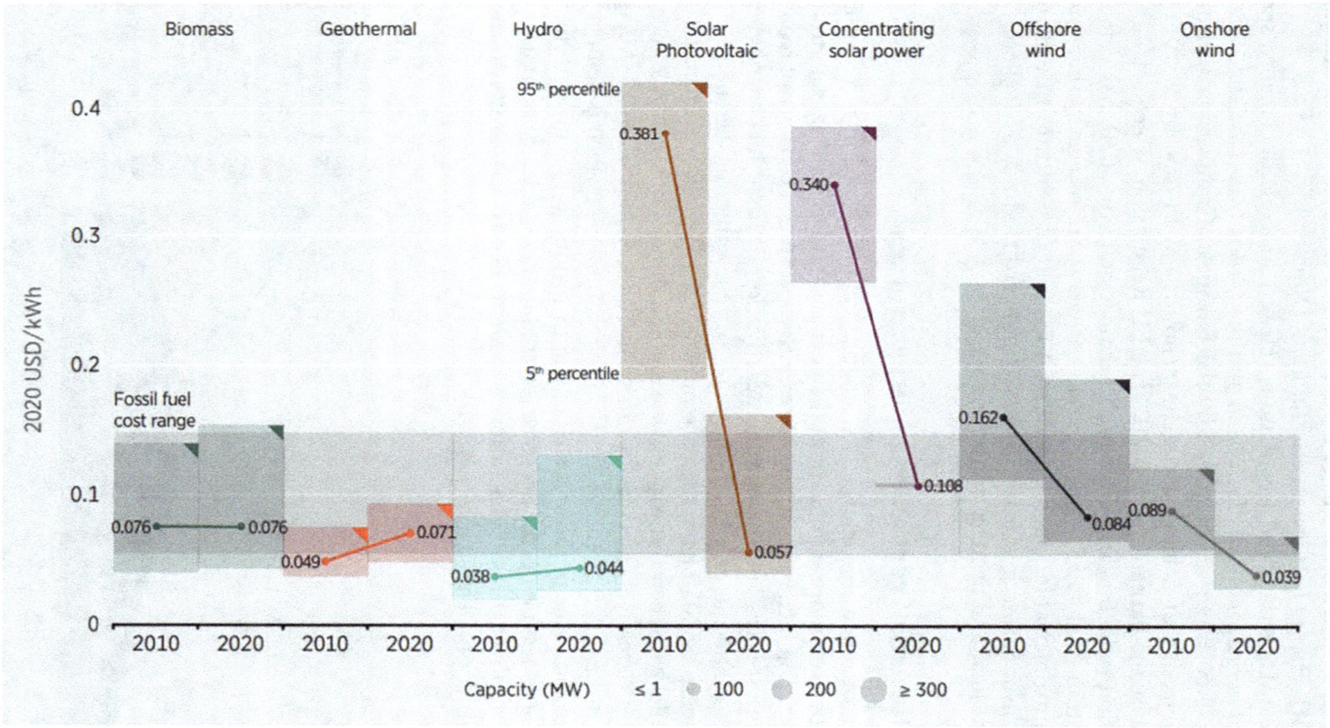

Figure 1.2 Global LCOE for new plants for various power generation technologies, 2010–2020.

International Energy Agency.

In the figure, the lines represent the global weighted average LCOE derived for each plant, which is calculated with a *Real Weighted Average Cost of Capital* (WACC) of 7.5% for OECD countries and China in 2010, declining to 5% in 2020, and 10% for the rest of the world in 2010, declining to 7.5% in 2020 (IRENA *Renewable Cost Database*). The horizontal band highlighted in orange in Figure 1.2 represents the cost variation of fossil fuel power generation, while the bands for each technology represent the 5th and 95th percentiles for plants using renewable sources.

Decarbonization is also expected to have a significant positive impact on the global economy: International Renewable Energy Agency (IRENA), in fact, estimates that this change in the energy sector could lead to an increase in global GDP (Gross domestic product) of more than 1% by 2030. In this regard, it is important to note that, over the years, mainly due to energy efficiency measures, a decoupling has been created between the level of pollution (defined in terms of CO_2 emissions) and energy intensity: this is an approximate indicator of an economy's energy efficiency and relates the quantity of energy consumed to the level of economic production, represented by GDP, and is defined as the ratio between gross domestic energy consumption (if referring to primary energy, the unit of measurement is MTEP, (million tons of energy equivalent), or kWh) and gross domestic product (measured in dollars or euros): CIL/GDP (unit of measurement kWh/€) (Eurostat).

The concept of energy intensity allows us to better understand the evolution and energy situation of an economic system, as well as providing an approximate measure of the energy efficiency of a country's production and consumption systems. Between 1990 and 2016, the EU's energy intensity decreased by 36%, with an average annual decrease of 1.7%. The decoupling between energy intensity and energy consumption is progressively demonstrated by the widening of the gap over the years. This trend is also evident when considering, in the relationship with GDP, the value of emissions instead of energy intensity.

A country's energy intensity is often used as an indicator of energy efficiency because, at the aggregate level, it is a measure of the energy needed to meet the required energy services; however, it should be kept in mind that a relatively low energy intensity does not necessarily correlate with high energy efficiency. For example, a small country with a mild climate would have a lower intensity than a large industrial country with a cold climate, even if energy were used more efficiently in the latter. Similarly, trends toward lower intensity do not necessarily correspond to improvements in system efficiency.

Other factors also influence intensity levels: the structure of the economy (presence of *energy-intensive industries*), geographic characteristics (e.g., longer distances implying greater demand for the transportation sector), overall climate and weather conditions (variations in heating or cooling demand), and the local economy. The development of renewable energy sources requires a shared definition.

In February 2013, the IRENA introduced a definition of renewable energy into its statute, ratified by 108 members (107 states and the European Union): "*Renewable energy includes all forms of energy produced from renewable sources in a sustainable manner, including bioenergy, geothermal energy, hydropower, ocean energy, solar energy, and wind energy.*" The International Energy Agency (IEA) defines renewable energy resources as those "*derived from natural processes and replenished at a faster rate than they are consumed.*" The IEA includes the following sources among renewables: electricity and heat derived from solar, wind, ocean, hydropower, biomass, geothermal resources, biofuels, and hydrogen derived from renewable resources. These are effectively two different definitions, varying in the type of sources included, and any reference to their sustainability is implicit.

For the definition of the various forms of renewable sources, it is possible to refer to European and national regulations. In Italy, the definition of energy from renewable sources is given in art. 2 of Legislative Decree 28/2011 (implementation of Directive 2009/28/EC on the promotion of renewable energy) and subsequent amendments:

- *Energy from renewable sources*: energy from non-fossil renewable sources, namely wind, solar, aerothermal, geothermal, hydrothermal, oceanic, hydraulic, biomass, landfill gas, waste gas from sewage treatment processes, and biogas;
- *Aerothermal energy*: energy accumulated in the air in the form of heat;
- *Geothermal energy*: energy stored in the Earth's crust in the form of heat;
- *Hydrothermal energy*: energy stored in surface water in the form of heat.

In short, this definition includes all forms of non-fossil energy that are continuously regenerated: wind, solar, aerothermal, geothermal, hydrothermal, oceanic, and hydraulic. It also includes bioenergy sources such as biomass, biofuels, bioliquids, landfill gas, sewage treatment gas, and biogas.

Recalling sustainability principles is essential because energy production from these sources inevitably has both positive and negative environmental, economic, and social impacts, which must be appropriately and carefully assessed. These considerations are more relevant for bioenergy but are equally important for other technologies, especially if they are widely used. Concerns related to the sustainability of bioenergy production and use, for example, require a specific, case-by-case assessment based on the raw materials used, production methods and their efficiency, land use, infrastructure, and so on. These in-depth analyses must be conducted for all types of bioenergy, from traditional uses of biomass in the residential sector to bioenergy used in transportation and energy production.

The assessment of greenhouse gas (GHG) emissions must be conducted with an adequate assessment of the entire life cycle, from land use to end uses. The outcome may differ, for example, depending on how the indirect impacts of bioenergy production on land use (*Indirect Land Use Change*, ILUC) are considered or on the correct assessment of the energy generation system in terms of efficiency in the case of biomass. Furthermore, the potential economic and social impacts of bioenergy use, including food security, must be carefully assessed.

Many of the concerns surrounding bioenergy relate to its relationship with agriculture. To be truly sustainable, the agro-ecological model must consider the circularity of production: renewable energy can be obtained from *bioenergy* when other sources (solar and wind) are not producing, and vice versa. This model also inspires *agrivoltaics*, a smart way to combine agriculture with renewables. This system allows for the introduction of photovoltaic solar energy production on farms, integrating it with crop and livestock production: the panels, which can also be *mobile* and *solar-tracking*, are positioned in the fields with shapes, sizes, and heights that allow for agricultural work and grazing. This is an interesting form of integration, both for achieving the energy system's decarbonization goals and for the sustainability of the agricultural system and the profitability of farms. Involving the agricultural sector is crucial also to stimulate the reclamation of abandoned and currently unproductive agricultural land.

Agrivoltaics enables agricultural yields that would otherwise be unattainable and ensures that energy is produced without land consumption or polluting emissions into the atmosphere. Therefore, it is not a model that pits food production against energy production, but rather its ultimate objectives are multiple and synergistic: productive, economic, and environmental. Furthermore, farmers can also profit from energy production and land; even those currently disused are thus preserved and protected from extreme events due to climate change.

Another virtuous example in the bioenergy sector is biogas: a mixture of methane gas and carbon dioxide produced by the oxygen-free fermentation of organic matrices (anaerobic digestion). Sustainability in this case is achieved if the use of biogas contributes to improving the environmental impact of livestock effluents by transforming them into renewable fertilizers, if the byproducts of livestock farming and agro-industrial activities are valorized, and if agricultural technology innovation is stimulated, favoring the development of land management systems and local resources. A biogas plant, in fact, has the potential to increase the efficiency of crop rotations without hindering food production, to use digestate as a renewable fertilizer, and to use excess thermal energy to meet the farm's heating needs.

Agrivoltaics and the use of biogas, if used well, interpret the synergy between different sectors in the most correct way because they do not put the energy sector in competition with the food sector. It is in fact important to underline the social aspect of the new energy model that uses renewables: the integration of the world of consumption (in the latter case, consumption of food and energy) with that of production (of food and energy).

1.1.3 The digitalization of energy and distributed energy generation

Digitalization offers a series of essential advantages for the development of the new energy model; indeed, it is its driving force for several reasons: a more reliable, safe, and sustainable electricity grid to also increase resource utilization; intelligent control; the use of historical databases; and automatic optimization of the supply/demand system. Digitalization improves the safety, productivity, flexibility, accessibility, and sustainability of energy systems, although it introduces security and privacy concerns (IEA, 2021).

These new technological developments have enabled a shift in the traditional relationship between energy supply and demand: The presence of interconnected digital systems increases the flexibility of energy demand, aligning it with the typical supply of renewables, and allows for its aggregation into homogeneous packages, optimizing costs and quantities. On the supply side, it also enables a shift from centralized to decentralized generation, with obvious advantages in terms of yields and grid losses.

The progressive diffusion of distributed energy generation is directly linked to the increasing level of digitalization of the system; the high connectivity resulting from electrification and decentralization is transforming the way electricity is produced and consumed. The electrification of large sectors of the economy is one of the most important expectations of the energy transition, especially in the transport and heating sectors, where electrification is already a mature technology: the number of electric vehicles will play a significant role in this development.

Finally, digitalization will make the grid smarter by implementing new technologies, such as smart meters and sensors, or automation technologies, connecting devices with the Internet of Things (IoT): These new technologies create a more advanced and sophisticated network but also require adequate infrastructure.

The intensification of electricity demand, the increase in the share of variable renewable generation, and the integration of distributed generation systems are already addressed by smart grids, which improve the transmission and distribution network, optimize network monitoring, and enable the active contribution of the end user. A *smart grid* is able to implement all energy generation and storage options to optimize energy efficiency and ensure a fundamental feature of the new energy model: consumer activity.

The increased use of renewable sources for the production of renewable energy sources, such as renewable electricity, renewable gas, or green hydrogen, will require an increase in distributed generation. As energy becomes increasingly local, local market models based on cooperation and energy exchange between users, based on a logic of territorial proximity, will become increasingly important.

The role of individuals must adapt to proactive behavior, not only passive consumption but also production and responsibility for their own self-generated energy, according to the principles of energy communities. Europe has already led the way for member states to guarantee their citizens the right to produce renewable energy for their own consumption, to store it, and to sell excess production *without discrimination*.

1.1.4 The concept of energy communities

Energy communities play an important role in the development of new energy market models (renewable energy producers/consumers), including citizen energy communities or renewable energy communities. The latter, in particular, involve groups of citizens, social entrepreneurs, public authorities, and local entities, such as municipalities, who participate directly in the energy transition by investing in, producing, selling, and distributing renewable energy. In addition to reducing greenhouse gas emissions, communities offer many economic, social, and security benefits. The primary objective is to provide environmental, economic, or social benefits at the community level to all members of the local areas in which they operate, rather than financial profits.

The European Renewable Energy Directive requires a series of activities by member states to promote renewable energy communities:

- Ensure that final customers, in particular household customers, have the right to participate in a renewable energy producer/consumer community, while maintaining their rights or obligations as final customers and without being subject to unjustified or discriminatory conditions or procedures, provided that such participation does not constitute their principal commercial or professional activity;
- Ensure that renewable energy communities have the right to:
  - Produce, consume, store, and sell renewable energy;
  - Share renewable energy produced within the same community, while maintaining the rights and obligations of community members as end customers;
  - Access all appropriate electricity markets, directly or through aggregation, in a non-discriminatory manner;
- Provide a supporting framework to promote and facilitate the development of renewable energy communities based on the potential development of energy communities in their respective territories.

This framework ensures that participation in renewable energy communities is open to all consumers, including those from low-income or vulnerable households, and that all necessary tools must be made available to facilitate access to dedicated financing.

It's fair to say that renewable energy communities represent a way of producing, managing, and using energy that perfectly integrates the characteristics of the ongoing energy transition. For definitions of the entities and functions of a renewable energy community, see, for example, the report from the Italian Energy Authority (ARERA, 2020).

1.2 STRATEGIES FOR DECARBONIZATION

1.2.1 The Paris Agreement and its developments

The *Intergovernmental Panel on Climate Change* (IPCC), the international organization mandated by the United Nations and 195 countries to monitor the climate crisis and develop a strategy to combat it, has over the years painted a picture of an increasingly alarming crisis. First of all, none of the scenarios developed by the IPCC will be able to contain the temperature increase to 1.5°C by 2100: without immediate and far-reaching interventions on greenhouse gases, the 2°C threshold will also be exceeded, with serious repercussions on the intensification of extreme weather events, sea level rise, the water cycle, and ice melt. The amount of CO_2 that can still be emitted (*carbon budget*) before exceeding the 2°C limit is now very limited. Table 1.1 (IPCC, 2021) represents well the picture of a very critical situation for the planet, since the remaining carbon budget must be less than 300 Gt CO_2 to have a high probability of not exceeding +1.5°C, but at current rates, this value will be exceeded in 2028.

On the other hand, some of the effects of climate change have already become irreversible, at least from a human perspective: the melting of polar ice, Alpine glaciers, and permafrost, rising sea levels, and ocean warming and acidification are now structural variables of the Earth system that will be reckoned with for centuries. At the same time, extreme weather events are occurring with ever-increasing frequency and intensity. Overall, furthermore, the worsening of weather conditions will have dramatic consequences on the agricultural sector, particularly in the Mediterranean region (the most affected in the world, together with the West Coast of the USA).

The change in surface temperature (as an annual average) reported in Figure 1.3 shows man's responsibility and confirms the current increase of +1.1°C compared to the values before 1850. Compare the two trends in Figure 1.3, one observed, therefore real, and one simulated with the hypothesis of eliminating human intervention. The Paris Agreement is the first universal and legally binding agreement on climate change, adopted at the COP 21 climate conference in Paris in December 2015. They established a global framework to combat climate change, limiting global warming to below 2°C, a value considered the limit, and even trying not to exceed 1.5°C; thus, countries are required to implement concrete actions to address the impacts of climate change and to support them in their efforts, starting first of all from the evaluation of the average surface temperature of the planet compared to reference levels (1850).

Table 1.1 Temperature changes due to CO_2 emissions and remaining carbon budget

Global warming between 1850–1900 and 2010–2019 (°C)		*Historical cumulative CO_2 emissions from 1850 to 2019 (Gt CO_2)*					
1.07 (0.8–1.3; likely range)		2390 (±240; likely range)					
Approximate global warming relative to 1850-1900 until temperature limits (°C)	**Approximate global warming relative to 2010-2019 until temperature limits (°C)**	**Estimated remaining carbon budgets from the beginning of 2020 (Gt CO 2)/*Likelihood of limiting global warming to temperature limit*[a]**					**Variations in reductions in non-CO2 emissions[b]**
		17%	33%	50%	67%	83%	
1.5	0.43	900	650	500	400	300	Higher or lower reductions in accompanying non-CO_2 emissions can increase or decrease the values on the left by 220 $GtCO_2$ or more
1.7	0.63	1450	1050	850	700	550	
2.0	0.93	2300	1700	1350	1150	900	

Source: IPCC (2021).

a This likelihood is based on the uncertainty in transient climate response to cumulative CO_2 emissions (TCRE) and additional Earth system feedbacks and provides the probability that global warming will not exceed the temperature levels provided in the two left columns. Uncertainties related to historical warming (±500 Gt CO_2) and non-CO_2 forcing and response (±220 Gt CO_2) are partially addressed by the assessed uncertainty in TCRE, but uncertainties in recent emissions since 2015 (±20 Gt CO_2) and the climate response after net zero CO_2 emissions are reached (±420 Gt CO_2) are separate.

b Remaining carbon budget estimates consider the warming from non-CO_2 drivers as implied by the scenarios assessed in SR1.5. The Working Group III Contribution to AR6 will assess the mitigation of non-CO_2 emissions.

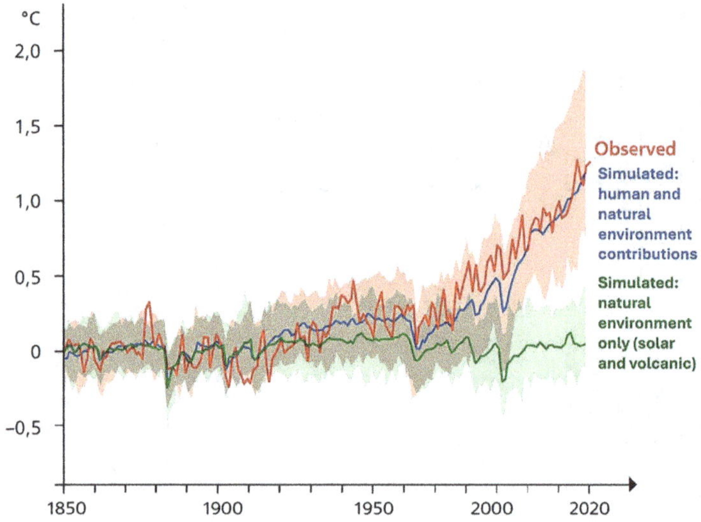

Figure 1.3 Change in average surface temperature (1850–2020).

IPCC, 2021.

The EU and its member states are among the 190 parties to the Paris Agreement. It formally ratified the agreement on October 5, 2016, thus allowing it to enter into force on November 4, 2016; for this to happen, at least 55 countries representing at least 55% of global emissions had to deposit their instruments of ratification. The Paris Agreement refers to a global strategy needed to achieve climate neutrality by 2050 in a context of mitigating the effects of climate change, focusing primarily on reducing emissions.

In this regard, the various states have agreed to:

- Keep the increase in global average temperature to well below 2°C above pre-industrial levels as a long-term goal;
- Limit the increase to 1.5°C, as this would significantly reduce the risks and impacts of climate change;
- Ensure that global emissions peak as soon as possible, while recognizing that developing countries will take longer to achieve this;
- Achieve rapid reductions thereafter, according to the best available scientific knowledge, so as to reach a balance between emissions and removals in the second half of the century.

To contribute to the agreement's goals, countries have submitted comprehensive national climate action plans, called *Nationally Determined Contributions* (NDCs). These are not yet sufficient to achieve the agreed temperature targets, but the agreement charts the course for further action. Among adaptation policies, governments have agreed to:

- Strengthen the capacity to address the impacts of climate change;
- Provide developing countries with continued and increased international support for adaptation.

The agreement recognizes the importance of averting, minimizing, and addressing loss and damage associated with the adverse effects of climate change, highlighting the important role of cities, regions, and local authorities; furthermore, it recognizes the role of stakeholders who are not parties to the agreement in addressing climate change, including cities, other subnational level entities, civil society, the private sector, and others.

Despite tangible evidence of human-caused climate change, widespread support for the Paris Agreement, and the prevalence of green, affordable, and sustainable energy options, energy-related CO_2 emissions still increased by 1.3% annually, on average, between 2014 and 2019. IRENA has identified a 1.5°C scenario (called 1.5-S) that describes an energy transition pathway to avoid exceeding the 1.5°C threshold, envisaging a set of technological solutions already available on the market. Time is of the essence, and a rapid decline in emissions must begin now: in line with the IPCC report on limiting global warming to 1.5°C by 2050, coal and oil are expected to have already peaked, while natural gas is expected to do so in 2025. IRENA charts a steep and continuing downward trajectory toward a 45% decline in carbon dioxide (CO_2) emissions by 2030 compared to 2010 levels, reaching net zero by 2050, in line with the IPCC agenda.

IRENA's World Energy Transitions Outlook is a path compatible with the 1.5-S scenario that also examines the socioeconomic and political implications, providing insights into structural and financial changes. Technologies for rapid decarbonization are increasingly available, but the energy transition is not confined to the energy aspect alone: in fact, energy systems based on renewables will produce profound changes that will impact all economies and societies. IRENA compares the 1.5-S scenario with the reference energy scenario (Planned Energy Scenario, PES), which provides a perspective on system developments based on governments' current plans (NDCs). In this context, the 2030 targets represent an important step in the journey: the choices made in the coming years will determine whether a 1.5°C future can be achieved. This perspective is in line with the 2030 Agenda of the United Nations for Sustainable Development and the SDGs, that is, its objectives.

IRENA's 1.5-S scenario actions provide guidance for:

- Follow the path most likely to reduce energy emissions over the next decade and put the world on a trajectory to limit temperature rise to 1.5°C;

- Support emerging technologies that are more likely to become competitive in the short term and are more effective in achieving long-term emissions reductions;
- Immediately limit investments in fossil fuels to facilitate their rapid replacement;
- Eliminate direct and indirect subsidies for fossil fuels;
- Adapt market structures to the new energy era;
- Invest in policies to protect communities affected by climate change and energy transition actions.

The necessary actions essentially concern renewable energy through direct supply, electrification, energy efficiency, and green hydrogen. There is no need to wait for new technical solutions; effective results can be achieved with existing options, but reaching the necessary levels quickly enough to meet the 1.5°C goal requires targeted policies and measures in a coherent transition. For now, the speed of the energy transition is far from that required to stay in line with the Paris Agreement. The policies currently in place, outlined in the *Planned Energy Scenario* (PES), would limit global emissions to stabilizing, with a slight decline toward 2050. Overall, the future projections indicated in the PES are far below what is needed to avoid exceeding the 1.5°C limit.

Overall, CO_2 emissions in this scenario in 2050 are projected to be –0.4 Gt C/year (billion tons of carbon dioxide per year).

In the PES scenario, annual emissions reach 36.5 Gt CO_2 in 2050, a value equal to the current one, which, as we have seen, is absolutely insufficient to combat climate change. The reductions are shown, sector by sector, on the right side of the graph: to reach the 1.5-S scenario, emissions would have to drop to zero in all sectors. In line with the IPCC reports on limiting global warming to no more than 1.5°C by 2050, the 1.5-S scenario allows for steeper decreases in the first ten years (up to 2030) and continues with a declining but less steep trajectory thereafter, reaching net zero by 2050. Since the energy sector is currently responsible for more than 80% of anthropogenic CO_2 emissions, it has a central role in the decarbonization process, especially by 2030.

The future of fossil fuel use is much debated: it is expected to decrease by more than 75% by 2050, but it is expected to still have a primary role, mainly in energy and, to some extent, also in industry, providing 19% of the primary energy supply in 2050, with the use of natural gas.

Other sector analysts, however, call for a complete abandonment of all fossil fuels, perhaps leaving only their use in industry for petrochemical products. All scenarios, however, agree on halving coal production by 2030 and gradually eliminating it by 2050. In 2050, natural gas would be used mainly in industrial processes, in the production of blue hydrogen (coupled with CCS), and in power plants; even for this point, this scenario would be too lax regarding the use of fossil fuels.

One fact, however, is that the *oil & gas industries* have very quickly begun to invest in green energy technologies as an important part of their business model. Staying below 1.5°C implies that current investments in fossil fuels must be virtually eliminated over the next 10–20 years. Early planning for the phase-out of fossil fuels would also demonstrate a just transition, consistent with climate objectives, capable of rapidly reallocating and creating new green jobs and services.

The *World Energy Transition Outlook* outlines how to allocate CO_2 emission reductions between energy sources and uses in the 1.5-S scenario. The energies rin-Renewables include renewable electricity generation sources and the direct use of renewable heat and biomass; energy efficiency relies on measures related to demand reduction and efficiency improvements, including structural changes and circular economy practices; electrification includes the direct use of renewable electricity in transportation and building heating applications; hydrogen and its derivatives include synthetic fuels and feedstocks. Also included in the figure are sectors such as CCS, which describes carbon capture and storage from fossil fuels, primarily in industry, and BECCS (Bioenergy with carbon capture and storage), which is a carbon removal measure using bioenergy. Greater caution should be exercised with these latter assumptions, given the severity of the climate situation and, consequently, the urgency of finding definitive strategic solutions. By 2050, electricity will be the main energy carrier, rising from 20% of total final energy consumption in 2018 to over 50% in 2050, as there will be a significant electrification of end uses in the transport and heating sectors, with radical changes affecting lifestyles and the direct participation of individuals.

The rate of improvement in annual energy intensity must increase to 3%, almost two and a half times the historical trend. This will make it possible to reduce the energy intensity of the global economy by more than 60% by 2050. The review of the policies needed for the main actions of the energy transition is reported in Table 1.2.

1.2.2 Energy communities

A fundamental role in the path toward decarbonization is played by energy communities, which are able to elicit collective responses from groups of people who share the same energy, social, and environmental needs.

The Renewable Energy Directive (RED II) and the Electricity Market Directive (EMD), promoted by the European Commission, Council, and Parliament as part of the Clean Energy for All Europeans Package, recognize and promote energy communities at an institutional level, establishing that *renewable energy self-consumers produce renewable electricity for their own consumption and may store or sell self-generated renewable electricity, provided that such activities do not constitute their primary commercial or professional activity*. Communities thus become legal entities whose

Table 1.2 Energy transition strategies and objectives

Technological avenue	*Objective*	*Recommendations*
Renewables (power and direct uses)	Deploy renewable energy in end uses	These policies include regulatory measures that create a market, as well as fiscal and financial incentives to make them more affordable and increase their cost competitiveness compared to fossil-fuel-based solutions
	Deploy renewable energy in power sector	The choice of instrument and its design should consider the nature of the solution (e.g., utility scale, distributed, off grid), the sector's level of development, the power system's organizational structure, and broader policy objectives
Energy conservation and efficiency	Increase energy conservation and efficiency in heating and cooling	Energy efficiency policies such as strict building codes, support for building retrofits, and appliance standards are critical for the energy transition in buildings and industrial processes
	Increase energy conservation in transport	Decarbonizing the transport sector, among other measures, requires a shift from energy-intensive modes to low-carbon modes
Electrification of end uses	Electrify heating and cooling	Targets for renewable power should consider the rising demand from the electrification of end uses, in line with long-term decarbonization objectives. Moreover, policies and power system design are needed to support electrification in achieving its potential for providing system flexibility
	Electrify transport	
Green hydrogen	Support the development of green hydrogen	An enabling policy framework should consider four key pillars: a national green hydrogen strategy, priority setting, guarantees of origin, and enabling policies
Sustainable bioenergy	Ensure the sustainable use of bioenergy	Renewable energy is not exempt from sustainability concerns. Some of these concerns include greenhouse gas emissions related to land-use change, impacts on air and water quality, and biodiversity

shareholders or members are natural persons, SMEs, or local authorities, including municipal administrations, and whose primary objective is to provide community-level environmental, economic, or social benefits to their shareholders or members or to the local areas in which they operate, rather than financial profits.

Other market players potentially involved include installers, ESCos, maintenance providers, financiers, etc., without necessarily becoming members of the communities. Not all communities operating in the energy market have profit as their primary goal; instead, they set themselves the primary objective of meeting environmental, economic, and social needs and, only ultimately, of making a profit.

The evolution, including regulatory, of the complexity of the configurations that renewable energy communities can assume will go hand in hand with the increasingly diversified energy skills that the members of these communities will need to possess. Therefore, it is desirable that European citizens become increasingly aware of their own behaviors and the opportunities available to them and be as involved as possible in the energy transition processes.

Within the European Union, member states have the option of granting renewable energy communities the right to manage the local distribution network in two ways: a physical model, which involves the community using its own network to exchange energy between members; or a *virtual model*, which involves using the public network and specifying how much energy is actually shared in each time interval. The Italian Energy Authority identifies the virtual model as the simplest and most effective for enabling and operationalizing collective self-consumption schemes and renewable energy communities.

In Italy, self-consumption currently follows a *one-to-one model*, meaning one production unit (PU) serves one consumption unit (UC, for example, the building's shared utilities in the case of condominium self-consumption). If we instead consider a *one-to-many* collective self-consumption model (one PU and multiple UCs), two configurations are possible:

1. **Physical self-consumption scheme**, which provides for a direct private connection between the generation plant(s) and domestic/common users, with a single access point (Point of Delivery, POD) to the public network (Figure 1.4);
2. **Virtual self-consumption scheme** (also called commercial or extended perimeter), which provides for the use of the public network for the exchange of energy between generation and consumption units (Figure 1.5).

Physical self-consumption involves a single POD exchange with the grid; the energy produced and self-consumed remains within the perimeter of the

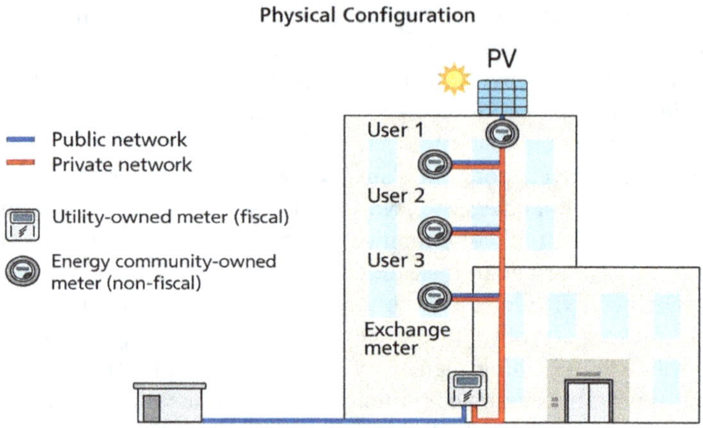

Figure 1.4 Physical self-consumption scheme with private connection of utilities to the production plant.

RSE.

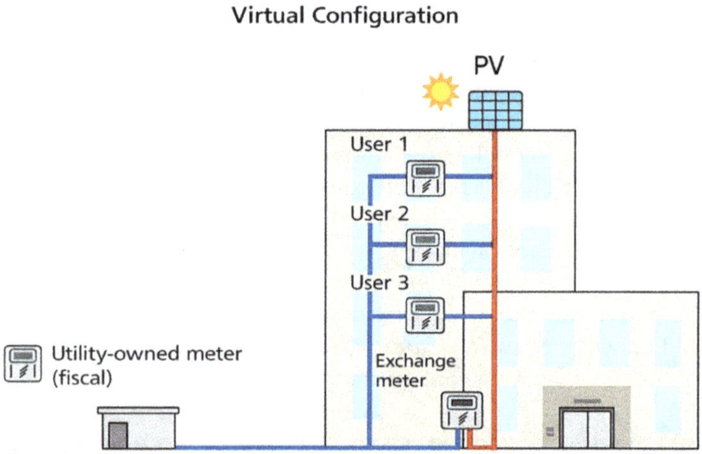

Figure 1.5 Virtual self-consumption scheme with public grid connection between users and the generation plant.

RSE.

building's private grid and is not subject to the variable portion of network and system charges.[2]

Virtual self-consumption requires that each user is connected to the public network via their own POD and is free to choose their own energy supplier or to break the pattern.

Physical energy exchanges continue to exist on portions of the public grid, which may be limited to the condominium's low-voltage busbar if the electricity meters are centralized in a dedicated room.

The methodology for calculating the distribution of virtual self-consumption is the result of contractual agreements between condominium owners and can be based on an energy criterion, for example, proportional to each user's consumption in each measurement period, or on a fixed criterion, such as a thousandths-based criterion, unrelated to the energy consumption of individual homes. It is important to consider the following general definitions:

- *Electricity actually fed in*: is the electrical energy fed into the grid net of conventional loss coefficients;
- *Electricity drawn*: is the electricity drawn from the grid by each user participating in the scheme;
- *Shared electricity for self-consumption* (or, more simply, shared electricity): it is, in each hour, the minimum between the sum of the electricity actually fed in and the sum of the electricity withdrawn via the connection points that are relevant for the purposes of a group of self-consumers of renewable energy acting collectively or of a renewable energy community.

Figure 1.6 briefly represents the energy flows relating to a condominium collective self-consumption scheme.

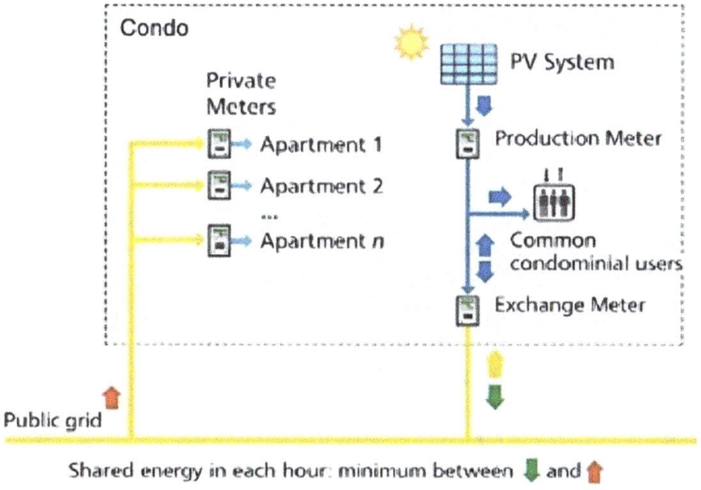

Figure 1.6 Energy flows of a collective self-consumption scheme: Energy produced, withdrawn, self-consumed, injected into the grid, and shared.

RSE.

For now, some experimental constraints have been placed on energy communities, but they will be updated and reduced over time:

- Renewable energy production plants must have a total power output not exceeding 200 kW;
- The production plants and withdrawal points forming part of a community must be connected to the low-voltage electricity grid through the same MT/BT transformation cabin (secondary cabin);
- Sharing of the energy produced with community members must take place through the existing electricity grid, including through storage systems;
- General system charges apply to energy drawn from the public grid, including shared grids;
- The incentive schemes for shared energy are decided by the Ministry of Economic Development (MiSE).

In particular, over the years, it will be possible to increase the installed power (at least up to 1 MW), connect the energy community to a primary substation, and talk not only about self-produced electricity but also about thermal energy.

1.2.3 Green hydrogen

Hydrogen and its derivatives will have to meet 15% of global final energy consumption by 2050. Its use will be crucial in energy-intensive and difficult-to-decarbonize sectors such as steel and chemical production, long-distance transport, maritime transport, and aviation. Hydrogen will also help balance supply and demand for renewable electricity and could represent an interesting long-term storage system.

The use of hydrogen is advantageous for many reasons: It is a carbon-free energy carrier; it can be produced using electricity, which, in turn, can be produced from a renewable source, such as wind or photovoltaic systems. In this case, we are talking about green hydrogen, the only option contemplated for future production. Its production can be delocalized, and surplus energy from renewable sources can be used, a practice included in the distributed generation process, which is beneficial for the self-sufficiency of less developed regions. As an energy carrier, it can distribute energy anywhere and in any season, and it is easily accumulated using tanks, by compression, or in liquid form, making it the most viable route for storing energy from renewable sources. Its high gravimetric energy density (kJ/kg) allows for seasonal storage, a clear advantage over electrochemical batteries, which cannot do so: in this way, it is possible to compensate, through its storage and subsequent release, for the non-programmability of renewable energy sources.

Unfortunately, its low density (kJ/m^3) makes long-distance transport via dedicated pipelines more challenging and expensive, but its distribution can utilize existing natural gas infrastructure. This important characteristic enables *sector coupling* between the electricity and gas sectors, through electricity generation from renewable sources, which is another key process for decarbonization. Hydrogen is therefore also a key element for this transition, given its ability to be mixed with natural gas and therefore transported through existing natural gas infrastructure.

Hydrogen is the most abundant chemical element in the universe but also the least used in terms of green energy; harnessing it would mean having a practically inexhaustible and completely natural energy source: it is found everywhere, in water and in living organisms, and it must be extracted from these to be used as an energy source, since it cannot be found in nature in the form of pure hydrogen.

The great challenge lies precisely in the difficulties inherent in its large-scale, affordable production. Hydrogen is produced primarily through a catalytic reaction between methane gas and high-temperature water vapor or by splitting water through electrolysis, producing electricity. The most widely used method for producing hydrogen today is the former, but natural gas is a fossil fuel, and this is absolutely unacceptable, as well as contradictory, from a decarbonization perspective.

Producing hydrogen with electricity from renewable sources, however, is the only method consistent with the challenges of the future, but currently it is 2–3 times more expensive than production from natural gas. Given the significant reduction in the costs of wind and photovoltaic power in recent years, it is hoped that the same process will also be achieved for hydrogen produced by electrolysis; for example, a significant advantage will be achieved by combining the electrolyzer section with the photovoltaic system.

Given that more than 95% of hydrogen is still produced using fossil fuels as a feedstock, industrial efforts should focus on the electrolysis process, which is the best option for producing hydrogen from non-fossil fuel feedstocks. Electrolyzers can be divided into three categories based on the nature of the electrolyte used: proton exchange membrane (PEM) electrolyzers, alkaline electrolyzers, and solid oxide electrolyzers (SOECs).

Finally, there is also the very interesting possibility of producing synthetic fuels through a reaction between hydrogen and CO_2: this system, if the hydrogen is produced from renewable sources, allows the absorption of significant quantities of CO_2 from the environment for the production of biogas or biomethane.

The EU's targets are highly challenging: hydrogen's share of the European energy mix should grow from the current 2 to 25% by 2050. To this end, at least 6 GW of electrolyzers will be installed in Europe between 2020 and 2024, producing up to 1 million tons of renewable hydrogen. These numbers will increase between 2025 and 2030, with at least 40 gigawatts of electrolyzers and the production of up to 10 million tons of renewable hydrogen. These prospects underscore the fact that hydrogen will already play a significant role in the integrated energy system over the next decade.

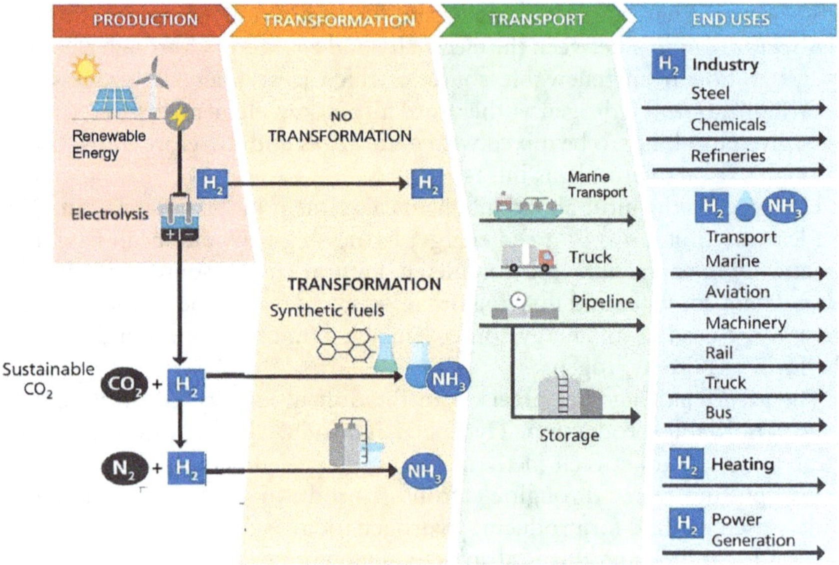

Figure 1.7 The value chain for green hydrogen in its various phases.

IRENA.

The great commitment of the nations toward the development of the hydrogen carrier must include a common strategy, which organically considers the various phases of this development: production, transformation, transport, and final use (Figure 1.7). This is why the so-called *hydrogen valleys* are considered privileged locations for technology transfer, which are geographical areas where various hydrogen applications are combined together in an integrated ecosystem, which includes production, consumption, experimentation, and training regarding the hydrogen carrier; these have a purely territorial character, refer to the use of hydrogen in the vicinity of its production site, and play an important role, in the intent of the European Hydrogen Strategy, in the propulsive activity toward the achievement of research and development objectives on a territorial level for the promotion of local industry.

1.2.4 Elements for the preparation of energy balances

The decarbonization process requires ongoing review of its consistency with energy production and end-use objectives, coordinating actions to utilize renewables, energy efficiency, and energy saving. The main global and national energy data on energy consumption and production are provided by the IEA, the United Nations, and the World Health Organization (WHO).

A basic scheme for defining the elements of an energy balance is summarily represented in Figure 1.8, which refers to primary energy and subsequent transformations and integrations up to the final requirement, in this case relating to energy for the needs of a building. This tool, in short, allows for

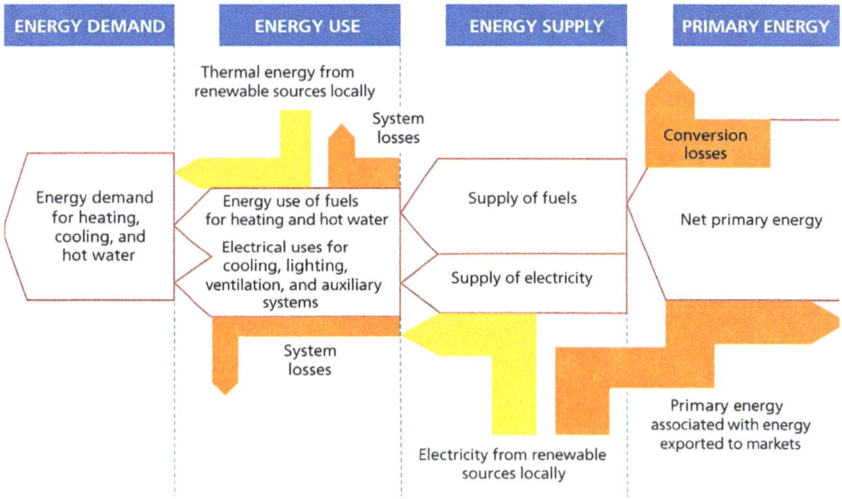

Figure 1.8 Schematic representation of the calculation of energy transformations, case study of a building (Commission Delegate Regulation (EU) No. 244/2012).

Delegate Regulation Commission.

a comparison of the role of various sources, moving from final energy demand by source and sector to demand by sector and use (heat, transportation, electricity, etc.).

Energy balance is particularly important for pursuing policies to rationalize energy use and the use of renewable sources. From the final energy demand per use, we move on to the evaluation of useful energy, that is, the energy that corresponds exactly to the service provided to the user and which completes the cycle from source to user. This analysis, which allows us to evaluate the relationship between useful energy and that used by energy-using devices, is of great importance for assessing the potential for energy saving and rationalization measures. Along the path between primary energy and useful energy, losses occur, which while not entirely eliminated, can at least be reduced, with significant economic and environmental benefits.

In the transformation phase from primary to secondary sources (especially in thermoelectric production), there are other losses, the extent of which is determined by the type of process.

Primary energy sources are defined as energy products extracted or obtained directly from natural resources, while those resulting from a transformation of primary energy sources are defined as secondary energy sources. To compare the available energy produced from different sources, it is necessary to refer to primary energy.

Primary energy is calculated using the primary energy conversion factor, a dimensionless ratio indicating the amount of primary energy for a given energy vector used to produce one unit of energy supplied. The factor takes into account the energy required for extraction, processing, storage, and transportation; and, in the case of electricity, the average efficiency of the generation system and the average transmission losses of the national electricity system; and, in the case of district heating, the average distribution losses of the grid.

Some categories of renewable energy production are not fully or consistently represented in the data, and for this reason, their energy balance must be carefully calculated. For example, small-scale distributed grid-connected energy generation, such as small-scale photovoltaics (1 MW), renewable energy production estimated based on installed capacity rather than field measurements, biofuels, which are currently measured at final rather than primary energy levels, thermal energy production from solar collectors, waste-to-energy (for which renewable waste fractions are not consistently distinguished), and assessments of energy produced by heat pumps, especially if renewable, are examples that demonstrate the need for specific in-depth analyses to achieve homogenization. Passive solar energy, which contributes to energy needs in industrial processes and building air conditioning, is typically overlooked due to the lack of universally shared calculation methods.

To identify the share of renewables in the global energy mix, it is necessary to define at which level of the energy balance the assessment should be carried out: primary energy or final energy. The advantages and disadvantages of using primary and final energy as the basis for monitoring are summarized in Table 1.3.

In the near future, addressing and solving the decarbonization process will mean continuing to develop policies that provide increasingly effective

Table 1.3 Methodologies for calculating primary energy supplied and final energy consumption

	Primary energy supply	*Final energy consumption*
Advantages	Widely used	Heat and electricity in a form ready for consumption
	Based on the physical measurement of fuels	Closer to useful energy output valued by end-users
		Better balance for directly produced RE
Disadvantages	Different conventions for assumptions on efficiencies mean that the contribution of RE depends on calculation procedure	Losses need to be allocated
	Underrepresents directly produced RE	

Table 1.4 Methodological challenges in renewable energy analysis

Challenge	*Proposed approach*
Definition of renewable energy	Energy from natural sources that are replenished at a faster rate than they are consumed, including hydro, bioenergy, geothermal, aerothermal, solar, wind, and ocean energy
Sustainability of renewable energy	Develop sustainability protocols for different forms of renewable energy over time, so that sustainability considerations can be incorporated into the definition in the medium term
Primary versus final energy accounting	Track renewable energy as a share of total final energy consumption, and as a subsidiary indicator, the share of renewable energy in electricity generation
Measuring additional indicators	Track complementary indicators such as deployment diversification, renewable energy policy, technology cost, and diversification

solutions to the major energy problems that are still in search of a definitive resolution. These challenges are summarized in Table 1.4 and highlight the long road that still needs to be followed to complete a transition full of pitfalls and obstacles.

1.2.5 Tackling energy poverty

The term energy poverty covers a broad spectrum of negative impacts on people's lives caused by problems with zero or insufficient energy access. In a situation of energy poverty, a family is unable to guarantee a level and quality of energy services sufficient to meet their social and material needs. This situation is often accompanied by the use of highly polluting fuels (e.g., kerosene or firewood), which leads to further health problems.

There is still no globally accepted definition of energy poverty, and without a unified definition, the number of families living in this situation cannot be adequately quantified, but at least statistical data describing some of the effects of energy poverty can help clarify the issue. Energy poverty exists in every part of the world. In the Western Balkans, for example, one in three families is unable to heat themselves adequately: families with incomes below 60% of the national median are particularly at risk (Energy Poverty observatory); by comparison, less than one in ten households in the European Union face the same challenge (but up to two in ten among poor households). Estimates indicate that 3 million households (out of 5 million) in the Western Balkans use obsolete wood-fired appliances, a major indicator of energy poverty. High dependence on biomass and poor regulatory mechanisms also endanger local forest resources, subject to illegal logging and black market sales. Governments in the region have taken steps to address energy vulnerability, first defining who "vulnerable customers" are: many provide some form of assistance to this population, mainly in the form of cross-subsidy, energy bill discounts, and protection from disconnection.

One of the sustainable development goals of the UN Agenda 2030 is to guarantee access to energy for all by 2030. In developing countries, according to the International Energy Agency, energy poverty affects approximately one billion people who lack access to electricity and approximately 2.7 billion people who use dirty and polluting fuels for cooking; however, this problem does not only concern the poorest countries because, for example, in Europe, the numbers are equally significant: 50 million people, of which approximately 5 million are in Italy.

According to the European Energy Poverty Observatory (EPOV) in Italy, the situation is particularly critical, with a much higher percentage of the population (16%) unable to adequately heat their homes than the European average (8.7%). In 2019, over two million families were experiencing energy poverty, equal to 8.5% of all families, based on a definition provided by the 2017 National Energy Strategy and included in the 2019 National Integrated Energy and Climate Plan (PNIEC), which also set the goal of reducing energy poverty to between 7 and 8% of all families by 2030. According to this definition, energy poverty is the difficulty in purchasing a minimum basket of energy goods and services or, alternatively, access to energy services that involves a diversion of resources, in terms of expenditure or income, above a "normal value." In practice, a person is in energy poverty if their equivalent energy expenditure is more than double the average household expenditure and if, at the same time, their total expenditure, net of energy expenditure, is below the relative poverty threshold, as identified by ISTAT (Istituto nazionale di statistica). Therefore, even a family with a total equivalent expenditure below the median, even if it has zero heating expenditure, is in energy poverty.

We could say that *energy poverty* is comparable to *absolute poverty*, considering the importance of access to energy for a healthy life, free from social hardship and deprivation that affects the very social dignity of individuals: a factor that concerns not only protection from the cold, but also from the heat and which is exacerbated by the effects of climate change.

The phenomenon has increased significantly over the years, especially among the most vulnerable groups, namely the young and the elderly, thus accentuating the income gap between generations and within the same generation; furthermore, it represents a social injustice, linked to an imbalance that has now become unacceptable between rich and poor (in Italy, in mid-2017, the richest 20% of Italians held over 66% of the national wealth while the poorest 50% owned only 8.5%), with intergenerational mobility virtually at a standstill.

The social disparity between individuals that is being experienced in these times is exacerbated by the effects of climate change and the social measures adopted to combat the COVID-19 pandemic, which often cause a condition of loneliness in contemporary man, which is also experienced as if it were a real exclusion, which is often a consequence of the lack of digital access to social media and the Internet.

Furthermore, digital analysis of societal transformations and, more generally, access to information is inhibited among the poorest segments of the population, and this is in any case due to the lack of access to energy: energy poverty effectively implies *social poverty*.

This is a complex problem, requiring a multidisciplinary and comprehensive approach to resolve. Unfortunately, current measures to combat it remain confined to welfare, which offers partial, objectively ineffective solutions. Energy expenditure must be reduced through structural interventions to improve the energy efficiency of homes, and energy costs must be controlled, again through structural interventions, with particular attention to the most vulnerable social groups.

Finally, there is a legislative problem, which can only be resolved through full recognition of energy as a fundamental social right. Facilitated access to energy, institutionally considered a common good, would solve the problem of energy poverty and would contribute more generally to reducing the negative effects of climate change. Countermeasures should also be funded, which require structural proposals, possibly of a fiscal nature (De Santoli, 2021).

1.3 ENERGY EFFICIENCY OF BUILDINGS

1.3.1 NZEB and PEB

European energy policies (Directives 2010/31/EU and 2018/844/EU) have introduced the *nearly zero energy building* (NZEB) objective to stimulate the energy transition of the construction sector, which alone accounts for 40% of final energy consumption, 27% of which is attributable to the residential sector, and which still largely uses fossil fuels (80%). Based on the most recent developments, the design models of NZEBs and their evolutions, *Positive Energy Buildings* (PEB), include innovative technologies and materials to reduce a building's energy needs, such as *Phase Change Materials* (PCM), *Vacuum Insulation Panels* (VIP), *smart glasses*, and integrated photovoltaic systems. Careful and integrated design of the building envelope and systems can not only allow for total self-production of energy consumption from renewable sources but also for generating an energy surplus to be shared with local urban networks and therefore assigning each building the role of a potential energy producer.

The 2010 EU Directive introduced an initial definition for NZEBs: buildings with high energy performance, meaning very low energy demand that is met to a significant extent by renewable sources, produced on-site or nearby. The application of the NZEB model in Europe became mandatory on December 31, 2018, for new public buildings and on December 31,

2020, for all new buildings. Furthermore, in order to achieve the Union's decarbonization and energy efficiency objectives, including to reduce the costs of non-ETS (*Emissions Trading Scheme*) systems (such as construction and transport), these models have also been extended to existing buildings undergoing major renovation. Decarbonization strategies for the building sector require the use of automation and control technologies for buildings and the introduction of smart grids for the sharing of renewable and distributed energy. The PEB can be considered different from a net-ZEB, which in some periods produces excess energy that is subsequently consumed, because it should instead produce excess energy throughout the year. This model, therefore, should allow covering the needs of other existing buildings in the suburban area to achieve energy independence.

The energy performance of a building must be determined on the basis of calculated data or actual consumption data relating to heating and cooling needs of rooms, the production of domestic hot water, ventilation, lighting, and electricity for other uses. It is expressed with a parameter, the primary energy (PE) index, which identifies the unit primary energy required, expressed in kWh m^{-2} $year^{-1}$, for the purposes of energy certification of the building in question to verify compliance with the minimum energy performance requirements imposed by current legislation. The methodology applied to determine the energy performance of a building must be continuously updated to follow the trend of innovation over time. Member States describe their calculation methodology following general European standards, such as ISO 52000-1, 52003-1, 52010-1, 52016-1, and 52018-1.

1.3.2 Energy performance of a building

The total global energy performance index ($EP_{gl,\,tot}$) is equal to the sum of the non-renewable ($EP_{gl,\,nren}$) and renewable ($EP_{gl,\,ren}$) energy; $EP_{gl,\,nren}$ is the value used to determine the building's energy class and indicates the fossil fuel PE consumed by the air-conditioned building per square meter of surface area each year. The energy class summarizes the building's energy quality and consumption, based on parameters that also depend on the building's location, shape (S/V ratio), orientation, and thermal characteristics.

In particular, the energy performance index indicates how much energy is consumed for the building to reach comfortable conditions and takes into account the need for non-renewable PE ($EP_{gl,\,nren}$) for winter and summer air conditioning, for the production of domestic hot water, for ventilation, and, in the case of the non-residential sector, for artificial lighting and the transport of people or goods.

Legal checks for minimum requirements relating to a building or its systems must be conducted with the $EP_{gl,\,tot}$ value. For correct energy certification, the building's energy performance certificate, in addition to identifying the overall energy performance index and the corresponding energy class, must also include information on the contributions of the individual energy services that contribute to determining it, in accordance with the mandatory minimum requirements for the characteristics of materials and systems. The performance ranges that identify the energy classes are obtained through multiplying coefficients that reduce/increase the $EP_{gl,\, nren}$ value: a low energy performance index corresponds to low energy consumption and therefore better energy performance, which also translates into a higher commercial value.

1.3.3 The evaluation of the cost-optimality for NZEBs

The European Union's objectives regarding the decarbonization of the building sector and the resulting spread of NZEBs require a design oriented toward energy-economic criteria, which includes the role of the energy consultant and economic expert within the designer. To achieve these objectives, the European Energy Efficiency Directive introduces the concept of cost-optimality, that is, the amount of energy required to meet the building's energy needs at the lowest cost over its estimated economic life cycle. This methodology consists of analyzing a building's potential maintenance and management costs, in addition to its construction, and extending this analysis to the entire life cycle of the materials and services used. The comparative methodological framework requires member states to:

- Define reference buildings characterized and representative for their functionality and climatic conditions; they can be residential or non-residential, new or existing;
- Define the energy efficiency measures assessed for the reference buildings, which may be measures for buildings as a whole, for building elements, or for a combination of building elements;
- Assess the final and PE needs of reference buildings with and without the proposed energy efficiency measures; the calculation method must be based on existing European standards;
- Calculate the costs (in terms of net present value, NPV) of energy efficiency measures over the expected economic life cycle applied to the reference buildings, taking into account investment, maintenance, and operating costs; earnings from the energy produced; and disposal costs.

Applying the cost-optimization methodology helps understand and manage all the implications of implementing the NZEBs requirements. If EU countries want to achieve the 2050 CO_2 reduction targets, the requirements for new buildings must be those of NZEB, but they must not neglect specific economic considerations for energy savings and emissions reductions associated with energy consumption. For example, an average value to consider could be 3 kg CO_2 m^{-2} $year^{-1}$,[3] therefore, the cost-optimization

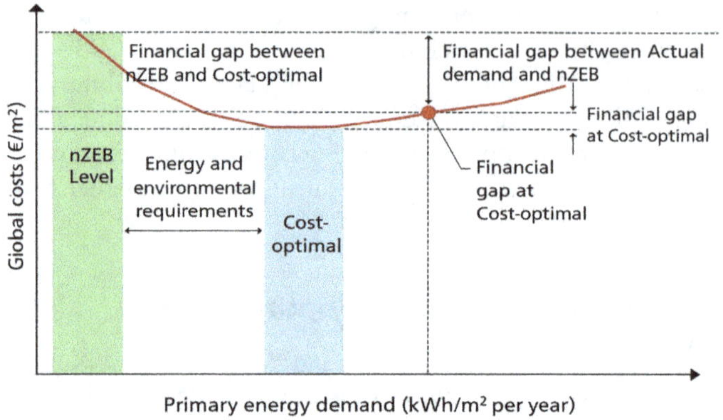

Figure 1.9 Financial, energy, and environmental gaps between existing building situations and those related to the cost-optimal analysis requirements and nZEB levels.

BPIE.

methodology must evaluate the cost of CO_2 emissions associated with the building's PE consumption. Furthermore, the related savings must be considered in the overall costs of a macroeconomic assessment, along with the savings resulting from energy efficiency measures.

The implementation of cost-optimality involves a selection of energy efficiency measures and policies for the building under consideration, organizing these solutions into intervention packages. These analyses consider the role of the building in its major sustainability aspects: energy, economic, and environmental. It will therefore be possible to identify three types of potential gaps to be progressively addressed to achieve decarbonization of the sector (Figure 1.9):

- *Financial gap*, that is, the actual cost difference between optimal levels as a function of the nZEB costs and levels;
- *Energy performance gap*, that is, the difference between the cost-optimal PE requirement and the nZEB levels;
- *Environmental gap*, that is, the difference between the CO_2 emissions associated with the demand for cost-optimal PE and nZEB levels, with carbon emissions expected to be nearly zero (or <3 kg CO_2 m^{-2} $year^{-1}$) to be consistent with the EU's 2050 decarbonization targets.

NOTES

1 When carbon dioxide is removed from the atmosphere, the terms carbon sequestration or carbon immobilization are commonly used. Net-zero emissions (climate neutrality or carbon neutrality) refer to achieving a balance between carbon emissions and carbon removals. Attaining this objective requires that greenhouse gas (GHG) emissions be offset by an equivalent amount of carbon uptake.

The primary natural carbon sinks include soils, forests, and oceans. Current estimates indicate that natural sinks remove approximately 10 Gt of CO_2 per year, whereas global CO_2 emissions are nearly four times higher, amounting to approximately 36–38 Gt annually.

2 The electricity bill includes, in addition to supply services (energy commodity costs, marketing, and retail services) and taxes, charges related to network services (transmission, distribution, and metering management), as well as components intended to cover costs associated with activities of general interest. These include, for example, financial support schemes for renewable energy sources (general system charges).

3 Starting from CO_2 emissions in the European building sector of approximately 1100 Mt CO_2 in 1990 (including both direct and indirect emissions related to space heating, domestic hot water production, and cooling), and assuming a total useful floor area of 38 billion m^2 by 2050, a 90% reduction in emissions would require average CO_2 emissions not exceeding 3 kg $CO_2/(m^2 \cdot year)$.

Specifically: 1100 Mt $CO_2 \times (100–90\%)/38$ billion $m^2 = 2.89$ kg $CO_2 \cdot m^{-2} \cdot year^{-1}$.

REFERENCES

Climate Change Report (IPCC), 2021.

Crutzen P. J., and Benvenuti nell'Antropocene, *L'uomo ha cambiato il clima, la Terra entra in unanuova era*, Mondadori, 2005.

De Santoli L., *Energia per la Gente*, Castelvecchi, 2021.

Energy Transition Outlook, 2020.

Friedlingstein P., et al., Global Carbon Budget 2020, *Earth System Science Data*, 12, 2020, pp. 3269–3340.

Osservatorio sulla povertà energetica, IRENA, 2020.

Report ARERA, 2020.

World Energy Outlook (IEA), 2021.

Chapter 2

Raw materials and recycling: Relevant SDGs: 12, 7, 8, 11, 13, 14, and 15

Silvia Serranti

2.1 DEFINITIONS AND FUNDAMENTAL CONCEPTS

2.1.1 Primary and secondary raw materials

Primary raw materials are materials or substances obtained through the exploitation of natural resources, which are subsequently subjected to a series of processing treatments aimed at producing goods. Primary raw materials may originate from *renewable resources*, of plant or animal origin, or from *non-renewable resources*, such as metallic and non-metallic minerals and fossil fuels. Renewable resources can only be defined as such if they are managed in a sustainable manner, that is, if their rate of consumption is lower than the rate at which they can be replenished without creating additional stress factors for the environment. Another way of classifying raw materials is the distinction between *energy raw materials*, which can be used for energy production and *non-energy raw materials*. Energy raw materials, in turn, may be either renewable or non-renewable: in the first case, for example, one refers to biomass; in the second case, to fossil fuels, namely coal, oil, and natural gas. Non-energy raw materials can also be either renewable or non-renewable: the former include agro-food raw materials, while the latter comprise minerals and rocks of different types extracted from quarries and mines.

To this category of "traditional" raw materials, a new type has been added more recently: *secondary raw materials*, which have gained increasing importance in the perspective of the circular economy and resource efficiency. Secondary raw materials are materials or substances obtained from the recycling of products of various types and origins that have reached the end of their life cycle, or from industrial processing residues, and which can be reintroduced into the production cycle of goods. They may be used either to obtain the same type of products from which they were originally derived or to produce different goods. In both cases, the term *raw material* refers to an economic concept. A raw material must, in fact, be characterized by specific quality requirements, which vary depending on its final use and which correspond, therefore, to market quotations. This aspect is particularly

DOI: 10.1201/9781003595700-3

relevant with regard to secondary raw materials, which must possess quality characteristics comparable to those of virgin raw materials that they are intended to replace.

2.1.2 Urban mining

The term *urban mining* refers to the set of actions (i.e., collection, characterization, recycling process, quality control, etc.) aimed at recovering secondary raw materials from waste generated in urban areas or, more generally, from the large stocks of materials incorporated in cities, such as buildings and infrastructures (Cossu, 2013; Cossu and Williams, 2015). The definition of *urban mining* was developed in analogy with the concept of *ore mining*, namely the extraction and processing of minerals from deposits in the Earth's crust to produce primary raw materials. Cities can, in fact, be regarded as actual mines from which raw materials can be extracted for the manufacturing of new products, starting from waste generated by all anthropogenic activities, both civil and industrial.

Buildings, infrastructures, industries, vehicles, electrical and electronic equipment, packaging, and many other products represent reserves of raw materials, with concentrations of useful, rare, or precious elements comparable to, or even higher than, those found in natural reserves. Landfills, too, can be exploited to recover raw materials: this process involves excavating the "layers" of waste placed in the landfills and subsequently treating them to recover materials such as metals, glass, and aggregates, as well as energy and reusable space within the site. This is known as *landfill mining*, one of the latest frontiers of scientific research for resource recovery.

Urban mining, therefore, represents a sustainable alternative to the exploitation of non-renewable natural resources, that is, to *ore mining*. For example, 1 ton of printed circuit boards from WEEE (waste from electrical and electronic equipment) contains on average more than 200 kg of copper, 46 kg of iron, nearly 28 kg of tin and aluminum, and about 18 kg of lead, along with smaller amounts of gold, silver, platinum, and other rare and valuable elements such as rare earths. In many cases, these concentrations exceed those found in mineral deposits.

The analogy between ore mining and urban mining is not only conceptual but also substantive, since all the technologies developed by humans over time for mineral processing have been transferred and adapted to waste treatment and, therefore, to the recycling sector. More specifically, the concepts, approaches, and activities commonly employed in the exploitation of mineral deposits, as well as in mineral processing at the plant scale, can also be applied both to the exploitation of "urban mines" and to the treatment of solid waste materials of different origins.

The industrial sector of waste treatment for the recovery of secondary raw materials developed more recently than mineral processing, which has much older origins. This has led to a technological transfer from mining engineering, particularly mineral processing methods (i.e., comminution, separation, etc.) to materials recycling: the methodologies applied to obtain raw materials from either minerals or waste produced in urban areas are, in most cases, the same. Both minerals and waste must undergo beneficiation processes to meet the quality standards required by the market. Beneficiation is achieved through the application of more or less complex operations that constitute the treatment process to transform a natural resource into a primary raw material or a waste into a secondary raw material. In both cases, the technical choices, aimed at maximizing raw material recovery, must be guided by accurate economic evaluations designed to ensure maximum profitability.

Table 2.1 provides a functional comparison between *ore mining* and *urban mining* systems across key domains of the resource supply chain, based on the conceptual framework proposed by Nakamura and Halada (2015). In *ore mining*, resource identification relies on geological exploration, while in *urban mining* this function is performed through the analysis of waste flows. At the system planning level, feasibility and mine planning in ore mining find their functional counterpart in the design of waste collection systems in urban mining. Resource recovery is then achieved through mining and mineral processing in the primary supply chain, whereas in urban mining the same function is fulfilled by waste collection and processing in dedicated treatment facilities aimed at material recovery and valorization. In both systems, the final output consists of raw materials supplied to downstream value chains, in the form of primary and secondary raw materials, respectively.

Urban mining, in addition to representing a sustainable solution for the efficient use of resources and the reduction of both non-renewable resource extraction and environmental pressures, also offers many advantages over *ore mining*. As already noted, in many cases the concentrations (or grades) of useful, rare, and precious elements are much higher in waste than in

Table 2.1 Functional comparison between ore mining and urban mining systems across key domains of the resource supply chain, including resource identification, system planning, recovery and processing, and material output.

	Resource identification	*System planning*	*Recovery and processing*	*Output*
Ore mining	Geological exploration	Feasibility and mine planning	Mining and mineral processing	Primary raw materials
Urban mining	Waste flow analysis	Waste collection system design	Waste collection and processing	Secondary raw materials

Source: Based on the conceptual framework proposed by Nakamura and Halada (2015). The table highlights functional equivalences between extraction-based and recovery-based resource supply systems.

mineral deposits. With reference to technological waste, for example, one ton of computers contains on average 200 g of gold and one ton of smartphones contains as much as 300 g; by comparison, a gold-bearing ore typically contains around 5 g/t of gold. Moreover, extracting minerals of interest from an ore deposit requires the handling of large volumes of rock, whereas electronic devices are available in millions of units. Finally, mineral deposits are located only in specific areas of the Earth's crust, while electronic waste is distributed much more evenly on a global scale.

2.1.3 The main categories of waste generated in urban areas

There are many urban resources from which secondary raw materials can be obtained, but four main categories can be identified (Figure 2.1):

- Waste from electrical and electronic equipment (WEEE)
- Municipal solid waste (MSW)
- Construction and demolition waste (CDW)
- End-of-life vehicles (ELVs)

Waste from electrical and electronic equipment (WEEE). They represent a mine of rare and precious metals and include computers, televisions, refrigerators, mobile phones, light bulbs, and so forth. It is one of the fastest-growing waste streams worldwide, but its recycling is often difficult due to its complex structure.

Municipal solid waste (MSW). They consist of household and commercial waste, comprising both organic and inorganic materials. This category includes packaging waste collected separately, such as glass, plastic, paper, aluminum, and steel, as well as the organic fraction, which is destined for composting and anaerobic digestion. The various fractions

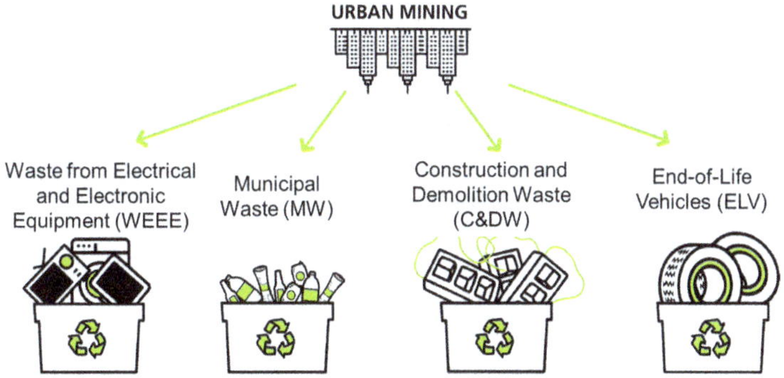

Figure 2.1 The main categories of waste streams representing the key resources of urban mines.

of collected waste are treated in dedicated plants to be recycled. The source separation carried out by consumers constitutes the first step in waste separation, and its quality is fundamental for the entire recycling chain. Once collected, post-consumer packaging is transported to sorting centers, where a preliminary separation is carried out, for example, the removal of bulky objects and other impurities, and the preparation of homogeneous material flows. The separation of materials is further refined in recycling facilities where washing and drying operations are also performed. The result is the production of secondary raw materials, destined for manufacturing companies to produce new goods.

Construction and demolition waste (CDW). This category includes all waste generated during the construction, maintenance, renovation, and demolition of buildings and infrastructures. Materials such as concrete, bricks, gypsum, plastics, wood, glass, and others can be recycled and transformed into secondary raw materials. The composition of CDW is highly variable, depending both on the diverse origin of the waste and on specific factors such as local construction techniques, climate, economic activities, and technological development in the area, as well as the local availability of raw materials and building products.

End-of-life vehicles (ELVs). They consist of a wide variety of components and materials, such as batteries, tires, oils, lamps, glass, plastic parts from bumpers and dashboards, seat fabrics and padding, and metal parts of the car body, making them rich in recoverable resources. An ELV is first depolluted, then dismantled to recover parts that can be directly reused as spare parts and, finally, subjected to material separation for recycling (plastics, glass, tires, etc.). The metal body is transferred to dedicated treatment facilities for the recycling of ferrous and non-ferrous metals. From an average ELV, weighing about 1.1 tons, treatment facilities can recover approximately 0.7–0.8 tons of ferrous metals and 60–80 kg of non-ferrous alloys. The overall recycling rate of a vehicle is around 85%. The fraction currently not recoverable is called car fluff, a heterogeneous mixture consisting of various materials (plastics, rubber, foam, fibers, small fragments of metals, and inert matter) that must be properly managed due to the potential release of contaminants into the environment. In some countries, car fluff can be used in the waste-to-energy plants; otherwise, it goes to a landfill.

In addition to the four main categories described above, there are many other urban resources that can be valorized for the production of secondary raw materials. For example, bottom ashes from waste-to-energy plants, from which aggregates and metals can be recovered; street sweeping waste, which yields primarily aggregates (i.e., sand and gravel), organic matter, and small amounts of metals; industrial sludges and slurries from which phosphorus, soil improvers, aggregates, and fibers can be obtained; and used oils, both mineral and vegetable, which are subjected to regeneration processes.

2.1.4 The circular economy

The currently predominant economic model, which originated and developed since the Industrial Revolution, is the linear one. According to this model, the life of a consumer product follows a "from cradle to grave" path consisting of the following steps: raw materials, production, use, and waste (Figure 2.2a). Raw materials are extracted and processed to manufacture objects that are purchased by consumers, who dispose of them once they no longer function or are no longer needed. This results not only in the indiscriminate and growing exploitation of raw materials, most of which are non-renewable resources destined to be depleted over time, but also in the increasing generation of waste, with significant environmental impacts on soil, water, and air. Moreover, the uncontrolled exploitation of natural resources (minerals, fossil fuels, and biomass) is exerting extreme pressure on the planet and is responsible for about 50% of all greenhouse gas emissions and more than 90% of biodiversity loss and water stress on a global scale (IRP, 2019).

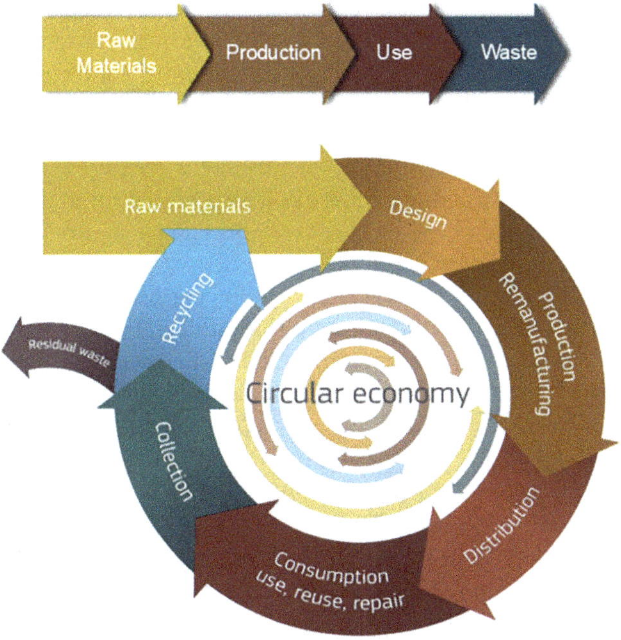

Figure 2.2 Comparison between the linear economy model and the circular economy model.

Source: EC, 2018a. © European Union, 2018. Reproduction is authorized provided the source is acknowledged.

It is, therefore, evident that the linear economy model is no longer sustainable and must be replaced by the new model of the circular economy (Figure 2.2b). This approach makes it possible to achieve a more efficient use of resources and a reduction in waste generation, with resulting environmental, economic, and social benefits.

Within the logic of the circular economy, waste disposal is minimized by extending the lifespan of products, promoting and encouraging their reuse or repair, and, when this is no longer possible, by recycling processes through which waste materials re-enter the production cycle, thus following a "from cradle to cradle" pathway.

One of the most widely recognized definitions of the circular economy was proposed by the Ellen MacArthur Foundation (2019), a leading global institution promoting and developing this concept:

> "...It is a generic term that describes an economy that is regenerative by design. In a circular economy, there are two types of material flows: biological ones, able to be reintegrated in the biosphere, and technical ones, destined to be re-used without ever entering the biosphere."

The circular economy model is inspired by the observation of nature, where even highly complex systems are organized in cyclical, regenerative schemes. In nature, the concept of waste does not exist: everything becomes nourishment for something else, for example, a leaf falling from a tree nourishes the forest. Similarly, in the circular economy, the "waste" of the linear economy can become nutrients for new production processes.

It is, therefore, possible to identify two different cycles to which the principles of the circular economy can be applied: a biological cycle and a technical cycle, represented in the well-known "butterfly diagram" proposed by the Ellen MacArthur Foundation (2019).

In the biological cycle, the so-called "biological nutrients" are involved, namely agricultural products, raw materials, and non-toxic organic waste that can be returned to the biosphere, sometimes also producing energy through processes such as composting, anaerobic digestion, and biochemical conversion. In the technical cycle, resources are represented by "technical nutrients," which make up many durable goods such as electrical and electronic equipment. Examples of technical nutrients include metals, plastics, and other non-biodegradable materials which cannot be absorbed by the biosphere but are destined to be transformed into valuable resources to be reused and recycled, thereby retaining their value.

The expansion of the circular economy will contribute significantly to achieve climate neutrality by 2050 and to decoupling economic growth from resource use, keeping the latter within planetary boundaries.

The transition from the linear to the circular economy model implies a profound and radical transformation involving all stages of the production chain: from product design, which must facilitate future disassembly (eco-design), to the development of new technologies and production processes, to product use, which must be extended as much as possible, and finally to waste recycling, which enables waste to be transformed into secondary raw materials to be reintroduced into the production cycle, replacing primary raw materials.

Table 2.2 lists, for each stage of the circular economy chain, the key actions needed to achieve the objectives of responsible production and consumption (EEA, 2016).

All links of the circular economy value chain are equally important; however, the last one, namely the transformation of waste into resources, represents a key element of the circular economy. To increase recycling rates and obtain high-quality secondary raw materials (*upcycling*), it is necessary first to improve waste management, invest in innovation in recycling processes, and create incentives to change consumer behavior. If we regenerate, reuse, and recycle, and if the waste of one sector becomes the raw material of another, we can move toward a more circular economy in which waste is recycled and resources are used efficiently and sustainably.

Table 2.2 Key actions to be undertaken at each stage of the circular economy value chain, aimed at achieving the objectives of responsible production and consumption

Raw materials	• Reduce the input of primary raw materials • Reduce material losses • Increase the share of secondary raw materials (from recycling) in total input • Use raw materials from sustainable sources
Eco-design	• Design products with longer lifespans • Design products to facilitate their disassembly (*design for disassembly*) • Incorporate recycled materials in product design • Design materials and products to be recyclable, minimizing losses in recycling processes
Production and distribution	• Use fewer materials in manufacturing • Reduce the volume and number of hazardous substances • Generate less waste • Shift business strategies toward circular concepts such as "*remanufacturing*"
Consumption	• Shift toward consumption models based on greener goods and services • Use products for longer • Generate less waste
Waste and recycling	• Increase recycling rates • Preserve material value in recycling processes, avoiding "*downcycling*" (i.e., the reduction in quality of secondary raw materials) • Optimize recycling system for both environmental and economic sustainability

Source: EEA, 2016.

Better waste management also helps reduce health and environmental problems, cut greenhouse gas emissions (both directly, by reducing emissions from landfills, and indirectly, by recycling materials that would otherwise be extracted and processed), and avoid local negative impacts such as landscape degradation caused by landfills, water and air pollution, and littering.

It must be noted that for the circular economy model to be truly effective, it must not only be implemented across all links of the value chain but also applied on a global scale, through cooperation with emerging and developing countries. The transition toward a circular economy is a complex process that requires significant changes in production and consumption patterns, including the financing mechanisms, consumer habits, government interventions in fiscal policies as well as technological, social, and business innovations.

2.2 EFFICIENT USE OF RESOURCES AND SDG 12

The management and efficient use of natural resources are of fundamental importance to ensure human well-being, environmental quality, and the growth of a sustainable economy. One of the main objectives (*Sustainable Development Goals – SDGs*) of the 2030 Agenda for Sustainable Development, namely *SDG 12 - Responsible Consumption and Production*, aims to achieve greater resource efficiency without halting economic growth, while at the same time reducing environmental degradation.

Over the past 50 years, raw material consumption has tripled and continues to grow (UNEP, 2024a), causing a continuous and progressive depletion of natural resources and severe negative impacts on air, water, and soil worldwide, as well as contributing to the growth of inequalities, both within individual countries and among different countries. On average, the richest countries consume 10 times more raw materials than the poorest ones, and global resource use is further increasing due to the rapid development of emerging economies.

To address this situation, SDG 12 promotes responsible consumption and production through efficient resource use in order not to jeopardize the needs of future generations while ensuring a better quality of life for all and the availability of decent and environmentally respectful jobs (United Nations, SDG 12).

Achieving this objective requires actions that lead to the provision of services and the production of goods that minimize the use of natural resources and toxic materials, while also reducing waste generation and pollutant emissions throughout the entire life cycle of the service or product. It is also necessary to act along the entire supply chain of a product, involving all actors from the producers to consumers. Consumer education on sustainable consumption and lifestyles, through information campaigns and appropriate tools, is a key aspect for achieving SDG 12.

A number of steps must be taken to improve the management of the different natural resources available on our planet. The consumption of non-renewable resources must be reduced by avoiding waste, using secondary raw materials derived from recycling processes, reducing the amount of raw materials needed to manufacture products, minimizing the use of toxic substances, and decreasing waste generation.

2.2.1 Global raw materials extraction

Raw materials play a central role in both economic development and human well-being (OECD, 2015). In many countries, the extraction and processing of raw materials, the manufacture of goods, and the management of end-of-life products are not only essential components of the economy but also provide significant sources of income and employment. However, the dependence on natural resources in production and consumption systems generates environmental, economic, and social consequences that extend far beyond national borders and will influence the prospects of future generations.

The way these resources are managed throughout their life cycle determines the rate at which renewable and non-renewable reserves are depleted, the magnitude of the environmental pressures generated during extraction, production, consumption, and disposal, and the resulting effects on air, water, soil, climate, biodiversity, landscapes, and human health. Resource management also shapes international trade dynamics and commodity prices, as well as economic performance and competitiveness.

Despite growing awareness of the environmental and health impacts associated with resource use, as well as the finite nature of non-renewable resources, global extraction and consumption of raw materials continue to rise. This trend is directly linked to what UNEP (2024a) refers to as the *triple planetary crisis*: climate change, biodiversity loss, and pollution and waste. Projections indicate that, under current patterns of resource use, global temperatures could rise by as much as 3°C by the end of the century (IRP, 2017).

Global extraction of raw materials has risen steadily since 1900. Total extraction grew from about 6 billion tons in 1900 (EC, 2018b) to 95.1 billion tons in 2020 and is expected to reach more than 160 billion tons by 2050 (UNEP, 2024a) if production and consumption models will not change. This increase is largely driven by an expected 20% growth in world population combined with a high rise in per capita resource use, associated with economic growth in emerging economies. Figure 2.3 illustrates global trends in the extraction of raw materials, disaggregated into four major categories over the period 1970–2024 (UNEP, 2024a). The first category, biomass,

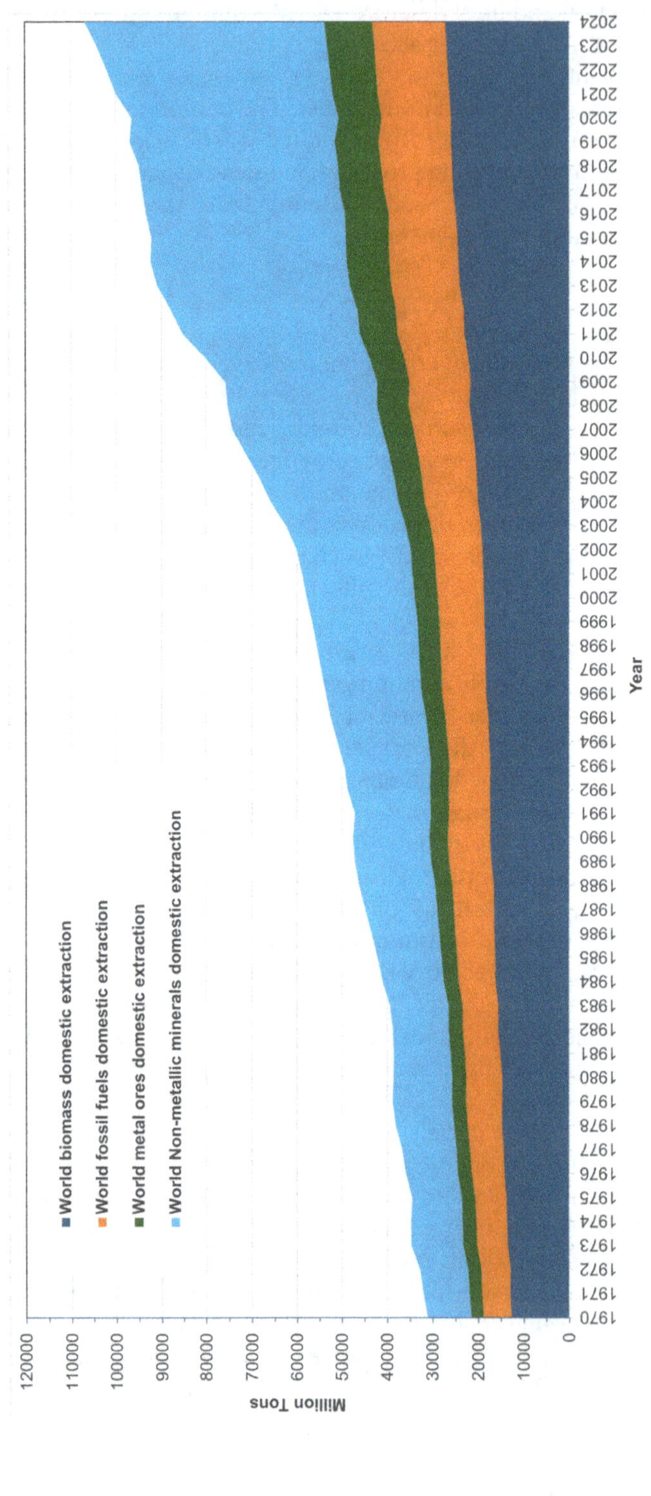

Figure 2.3 Global material extraction by major material categories (biomass, fossil fuels, metal ores, and non-metallic minerals), 1970–2024.

Data re-plotted by the author based on the United Nations Environment Programme, International Resource Panel, Global Material Flows Database (n.d.). Values originally reported in tons and converted to million tons (Mt). Data for the period 2020–2024 are projected estimates.

includes agricultural crops for food and energy production as well as wood for industrial and energy uses. The second, fossil fuels, comprises coal, oil, natural gas, and related resources. The third category refers to metal ores, including iron, copper, and aluminum, which are essential inputs for a wide range of manufacturing industries. Finally, the fourth category is related to non-metallic minerals, such as sand, gravel, clay, and other materials primarily employed in the construction sector. Non-metallic minerals show by far the steepest increase, overtaking biomass around 1970 as the dominant category, highlighting a profound transformation in the global socio-economic resource system, from one rooted in agriculture to one centered on construction materials and industrial demand. They are expected to remain the fastest-growing category, largely driven by infrastructure and construction demand, while biomass, fossil fuels, and metallic minerals will also continue to increase, though at different rates.

These figures highlight the urgent need to counteract current trends through measures aimed at improving resource efficiency and promoting sustainable development models based on the principles of the circular economy.

2.2.2 Changing geography of raw materials extraction and consumption

As illustrated in Figure 2.4, which presents raw material extraction by world region for selected years, the geography of global extraction has shifted markedly over the past decades from high-income countries to emerging and developing regions (UNEP, 2024a). In 1970, almost half of global extraction took place in Europe and North America while Asia and the Pacific, despite being the most populous region, accounted for a significant smaller share, followed by the former Soviet Union, Africa, Latin America, and the Caribbean and West Asia. By 2000, extraction had already increased considerably in Asia and the Pacific, reflecting accelerated industrialization and infrastructure development. This trend further intensified by 2019, when Asia and the Pacific clearly emerged as the dominant extraction region, while Europe and North America exhibited relatively limited growth, partly because these regions increasingly relied on imports from emerging economies. The most recent estimates for 2024 confirm this long-term reconfiguration, highlighting the structural shift of global extraction activities toward emerging regions.

While this expansion in resource extraction has contributed to poverty reduction and improved living standards in several countries, it has also generated severe environmental and social challenges. Moreover, a substantial share of the extracted materials has been exported for consumption in wealthier nations, leaving limited benefits for lower-income producer countries.

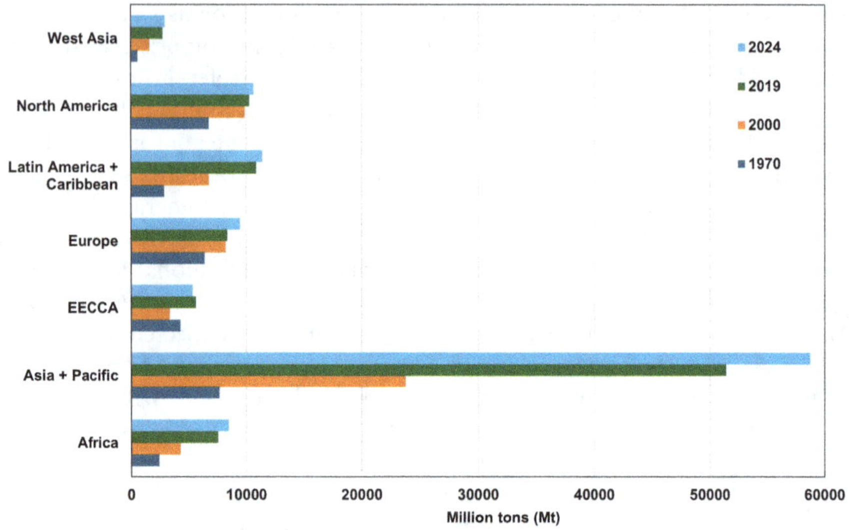

Figure 2.4 Raw material extraction by world region according to the UNEP classification for selected years (1970, 2000, 2019, and 2024).

Data for 1970–2019 are historical, while 2024 values are model-based projections. EECCA: Eastern Europe, Caucasus and Central Asia. Data re-plotted by the author based on the United Nations Environment Programme, International Resource Panel, Global Material Flows Database. Values originally reported in tons and converted to million tons (Mt).

The challenges are not limited to the depletion of non-renewable resources but also include the environmental impacts of extraction and processing activities. In 2022, the extraction and processing of biomass, minerals, and fossil fuels accounted for over 55% of global greenhouse gas emissions, up from about 50% in 2015 (IRP, 2019), contributed to up to 40% of particulate matter-related health impacts, and were responsible for over 90% of land-use-related biodiversity loss (UNEP, 2024a).

In addition to the rapid growth in raw material extraction, emerging economies are also experiencing a sharp rise in domestic material consumption, which translates into higher waste generation and environmental pressures (EC, 2018b). This makes the transition to resource efficiency particularly urgent in these regions.

As shown in Figure 2.5a, the trajectory of Asia stands out: material consumption has increased steeply in recent decades, surpassing that of the rest of the world around 2006. This trend has been driven primarily by China's rapid industrialization and urbanization, which demand vast amounts of materials such as steel and cement.

By contrast, as illustrated in Figure 2.5b, between 1970 and 2017, consumption levels in the EU-28 and North America remained relatively stable

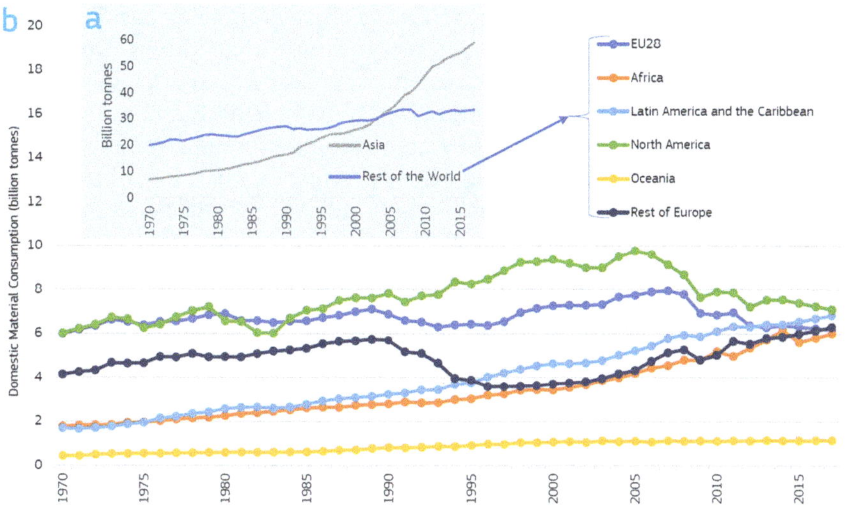

Figure 2.5 Domestic material consumption by region (1970–2017): (a) Material consumption in Asia compared to the rest of the world; (b) Material consumption in the EU-28, Africa, Latin America and the Caribbean, North America, Oceania, and other European countries.

Source: EC, 2018b. © European Union, 2018. Reproduction is authorized provided the source is acknowledged.

over the long term, ranging from 6 to 8 and 6 to 9 billion tons, respectively, while in the rest of the EU they fluctuated between 4 and 6 billion tons. Other world regions experienced more moderate but steady increases: Africa rose from about 2 to 6 billion tons, Latin America and the Caribbean from 2 to 8 billion tons, and Oceania from 0.4 to 1.1 billion tons.

These patterns suggest that future growth in global resource demand will be largely shaped by emerging and developing economies.

2.2.3 The concept of decoupling

The concept of decoupling is related to optimal resource efficiency, occurring when economic activity and human well-being continue to rise while resource consumption and its associated environmental impacts decline to levels that remain within planetary boundaries and are sustainable for future generations (IRP, 2019). This principle is often summarized as "doing better with less." Two forms of decoupling are usually distinguished: *relative decoupling*, when resource use grows more slowly than economic activity and improvements in well-being, and absolute *decoupling*, when environmental impacts decrease in absolute terms, while economic activity keeps rising (IRP, 2019).

Evidence since 2000 indicates that some relative decoupling has occurred, with environmental impacts rising more slowly than Gross Domestic Product (GDP) and the mass of extracted resources (IRP, 2019). However, in absolute terms, impacts are still increasing globally (UNEP, 2024a).

Achieving absolute decoupling in high-income economies could reduce average levels of resource consumption, promote a fairer distribution of prosperity, and help maintain high living standards. In contrast, in developing and transition economies, relative decoupling may still entail greater use of natural resources, but it can support rising income levels and poverty reduction, helping to secure socially acceptable living conditions (IRP, 2019).

Advancing toward decoupling requires ambitious and well-integrated policies, supported by scientific knowledge and technological innovation. This includes improving resource efficiency, fostering sustainable consumption, accelerating the transition toward a circular economy, and implementing measures for climate change mitigation and biodiversity protection (IRP, 2017). International cooperation can further accelerate the transition, creating fairer market conditions across countries, supporting decision-making in different contexts, and aligning efforts across countries with very different capacities and responsibilities.

The transition toward more efficient resource use relies on a combination of technological, social, and policy innovations alongside increased reuse and recycling, in order to keep resource extraction within safe ecological limits and align production and consumption with the Sustainable Development Goals (IRP, 2017). Within the UN 2030 Agenda, this is explicitly addressed by SDG 12, which focuses on ensuring responsible consumption and production (UN, 2015).

2.2.4 SDG 12: Targets, indicators, and relationships with other SDGs

SDG 12 on responsible consumption and production is supported by a framework of 11 targets and 13 indicators: the first represents the specific objectives, and the latter serves to monitor progress in their achievements (UN, 2015). The targets range from promoting sustainable management of natural resources, reducing food waste, and improving waste treatment to fostering sustainable corporate practices, green public procurement, and awareness on sustainable lifestyles. Table 2.3 lists these targets together with the corresponding indicators.

The importance of SDG 12 goes well beyond its own set of objectives, as responsible consumption and production patterns are closely intertwined with most of the other SDGs. SDG 12 can be seen as a balancing point between development, economic growth, and social well-being (SDGs 2, 3, 4, 6, 7, 8, 9, 11) on the one hand, and the safeguarding of natural resources and the planet (SDGs 13, 14, and 15) on the other. Table 2.4 illustrates some of the key synergies between SDG 12 and other goals. For instance, halving

Table 2.3 Official targets and indicators of SDG12 on responsible consumption and production

Target		*Indicators*	
12.1	Implement the 10-year framework of programs on sustainable consumption and production, all countries taking action, with developed countries taking the lead, taking into account the development and capabilities of developing countries.	12.1.1	Number of countries developing, adopting. or implementing policy instruments aimed at supporting the shift to sustainable consumption and production.
12.2	By 2030, achieve the sustainable management and efficient use of natural resources.	12.2.1 12.2.2	Material footprint, material footprint per capita, and material footprint per GDP. Domestic material consumption, domestic material consumption per capita, and domestic material consumption per GDP.
12.3	By 2030, halve per-capita global food waste at the retail and consumer levels and reduce food losses along production and supply chains, including post-harvest losses.	12.3.1	(*a*) Food loss index and (*b*) food waste index.
12.4	By 2020, achieve the environmentally sound management of chemicals and all wastes throughout their life cycle, in accordance with agreed international frameworks, and significantly reduce their release to air, water, and soil in order to minimize their adverse impacts on human health and the environment.	12.4.1 12.4.2	Number of parties to international multilateral environmental agreements on hazardous waste, and other chemicals that meet their commitments and obligations in transmitting information as required by each relevant agreement. (*a*) Hazardous waste generated per capita; and (*b*) proportion of hazardous waste treated, by type of treatment.
12.5	By 2030, substantially reduce waste generation through prevention, reduction, recycling, and reuse.	12.5.1	National recycling rate, tons of material recycled.
12.6	Encourage companies, especially large and transnational companies, to adopt sustainable practices, and to integrate sustainability information into their reporting cycle.	12.6.1	Number of companies publishing sustainability reports.

(*Continued*)

Table 2.3 (Continued)

12.7	Promote public procurement practices that are sustainable, in accordance with national policies and priorities.	12.7.1	Number of countries implementing sustainable public procurement policies and action plans.
12.8	By 2030, ensure that people everywhere have the relevant information and awareness for sustainable development and lifestyles in harmony with nature.	12.8.1	Extent to which (i) global citizenship education and (ii) education for sustainable development are mainstreamed in (*a*) national education policies; (*b*) curricula; (*c*) teacher education; and (*d*) student assessment.
12.a	Support developing countries to strengthen their scientific and technological capacity to move toward more sustainable patterns of consumption and production.	12.a.1	Installed renewable energy-generating capacity in developing and developed countries (in watts per capita).
12.b	Develop and implement tools to monitor sustainable development impacts for sustainable tourism that creates jobs and promotes local culture and products.	12.b.1	Implementation of standard accounting tools to monitor the economic and environmental aspects of tourism sustainability.
12.c	Rationalize inefficient fossil-fuel subsidies that encourage wasteful consumption by removing market distortions, in accordance with national circumstances, including by restructuring taxation and phasing out those harmful subsidies, where they exist, to reflect their environmental impacts, taking fully into account the specific needs and conditions of developing countries and minimizing the possible adverse impacts on their development in a manner that protects the poor and the affected communities.	12.c.1	Amount of fossil-fuel subsidies (production and consumption) per unit of GDP.

Source: UN (2015).

global food waste would be sufficient to eradicate hunger (SDG 2); improving resource efficiency would enhance water quality (SDG 6, target 6.3), support the transition to renewable energy (SDG 7, target 7.3) mitigate climate change (SDG 13), while reducing plastic and hazardous waste would

Table 2.4 Main SDGs directly connected with SDG 12

SDGs and main targets	*Correlation with SDG 12*
SDG 2 Ending hunger, achieving food security, improving nutrition, and promoting sustainable agriculture.	Halving food loss and waste would be sufficient to eradicate global hunger.
SDG 6 Ensuring the availability and sustainable management of water and sanitation for all.	Greater resource efficiency would contribute to improving water quality.
SDG 7 Ensuring access to affordable, reliable, sustainable, and modern energy systems for all.	Sustainable consumption and production models promote the use of renewable energy.
SDG 8 To promote lasting, inclusive, and sustainable economic growth, full and productive employment, and decent work for all.	Economic growth of countries and businesses is closely linked to achieving Goal 12.
SDG 9 Building resilient infrastructure and promoting innovation and fair, responsible and sustainable industrialization.	Sustainable use of natural resources is tied to essential needs such as water, food, and infrastructure.
SDG 11 Making cities and human settlements inclusive, safe, resilient, and sustainable.	More efficient use of natural resources contributes to healthier cities with lower environmental impacts.
SDG 13 Promoting action, at all levels, to combat climate change.	Sustainable production and consumption models must also ensure reductions in greenhouse gas emissions.
SDG 14 Conserve and sustainably use the oceans, seas, and marine resources for sustainable development.	Implementing Goal 12 can reduce human impacts on marine resources, both in terms of extraction and pollution.
SDG 15 Protecting, restoring, and promoting sustainable use of the terrestrial ecosystem.	Shaping production systems according to circular economy principles helps mitigate environmental impacts without hindering development.

directly benefit marine and terrestrial ecosystems (SDG 14, target 14.1 and SDG 15, target 15.1 and 15.5). Finally, education of the population (SDG 4) and the economic growth of countries and enterprises (SDG 8, target 8.4) are also closely linked to the implementation of SDG 12. SDG 12 is strongly connected with SDG 8 on sustainable economic growth and decent work, as both emphasize the need to decouple well-being from resource use and environmental degradation.

2.3 CRITICAL RAW MATERIALS

2.3.1 Definition, description, and main global producers

Raw materials are not only indispensable for daily life but also for enabling innovation in key strategic fields, making them a cornerstone of both economic development and industrial competitiveness.

The need for access to raw materials, not only in ever greater quantities but also in an increasing variety, has grown over the centuries in parallel with the scientific and technological progress. Over the past three decades in particular, the use of strategic elements has expanded dramatically to an extent never seen before.

The consumption of raw materials, especially metals, has risen exponentially across centuries to enable the production of increasingly efficient products and devices, capable of performing more complex and sophisticated functions while becoming progressively smaller through miniaturization. A clear example is the smartphone, a small device that can contain up to 50 different metals, almost half of the elements in the periodic table, required to achieve its compact size, light weight, and wide range of functionalities.

Ensuring a secure and uninterrupted supply of raw materials has become an issue of growing global concern. Demand is driven by multiple factors, including global population growth, industrialization, digitalization, and, more recently, the transition to climate neutrality. The latter, in particular, relies heavily on strategic raw materials required for low-carbon technologies and products, such as solar panels, wind turbines, electric vehicles, and energy-efficient lighting (EC, 2020a).

Concerns are further heightened by the fact that mining and refining activities are currently concentrated in a small number of countries. China, in particular, dominates global production, accounting for more than half of many critical minerals.

In the *European Green Deal* Communication of 11 December 2019 (COM, 2019), the European Commission explicitly recognized the supply of raw materials as a matter of strategic security for achieving climate neutrality by 2050.

In order to face this growing issue, the European Commission introduced in 2011 a list of critical raw materials (CRMs), regularly updated every 3 years.

A raw material is defined as critical on the basis of two main factors (EC, 2023a):

1. Economic importance (EI): calculated by assessing the role of a given raw material in the EU end-use applications and weighing it according to the relative economic value of these sectors that depend on it. The indicator also considers the technical and economic performance of possible substitutes in those applications.
2. Supply risk (SI): estimated by considering the likelihood of supply disruptions, taking into account factors such as global production concentration, import dependence, political and economic stability of producing countries, the availability of substitutes, and the recycling rates.

To compile the EU list of CRMs, a wide range of raw materials is evaluated, and both parameters are calculated. If the values for a given raw material exceed the established thresholds (EI ≥ 2.8 and SR ≥ 1), it is classified as critical.

The first list identified 14 CRMs; the number rose to 20 in 2014, 27 in 2017, and 30 in 2020. The most recent and fifth update, released in 2023, further expanded the list to 34 CRMs (EC, 2023a), reflecting the increasing strategic importance of raw materials in the EU's policy agenda (Table 2.5). A novelty introduced in 2023 is the definition of strategic raw materials

Table 2.5 List of the 34 CRMs identified by the EC in 2023, with SRMs in italics

Aluminum/Bauxite	*Copper*[a]	*LREE*	Scandium
Antimony	Feldspar	*Magnesium*	*Silicon metal*
Arsenic	Fluorspar	*Manganese*	Strontium
Baryte	*Gallium*	*Natural graphite*	Tantalum
Beryllium	*Germanium*	*Nickel*[a]	*Titanium metal*
Bismuth	Hafnium	Niobium	*Tungsten*
Boron/borates	Helium	*PGM*	Vanadium
Cobalt	*HREE*	Phosphate rock	
Coking coal	*Lithium*	Phosphorus	

Source: EC (2023a). Note: HREE: Heavy Rare Earth Elements; LREE: Light Rare Earth Elements; PGM: Platinum Group Metals

[a] Copper and nickel do not meet the critical raw material thresholds but are included as strategic raw materials under the EU Critical Raw Materials Act (EC, 2023b).

(SRMs). These are raw materials that, regardless of whether they meet the strict thresholds for criticality, are considered essential for the green and digital transitions, for defense, and for space applications. Copper and nickel, for example, have been added as SRMs given their importance for batteries, renewable energy technologies, and electrification, despite not being classified as critical under the EI-SR criteria. Among the 34 CRMs identified by the EC in 2023, 16 are classified as strategic, with heavy and light rare earth elements (HREEs and LREEs) considered together as a single group.

In the 2023 assessment, the position of each raw material in relation to the economic importance and supply risk thresholds is displayed in the EI-SR matrix (Figure 2.6). Materials located above both thresholds are classified as CRMs, while those also designated as SRMs are highlighted to reflect their additional strategic relevance.

Figure 2.7 shows the global map of countries that produce the largest shares of each of the 34 CRMs. China clearly dominates, as it accounts for the highest production volumes of many CRMs, most notably both light and heavy rare earth elements (REEs). Other key global suppliers include the Democratic Republic of Congo for cobalt, Russia and South Africa for PGMs, Australia for lithium, the United States for beryllium, and Brazil for niobium.

2.3.2 Strategic importance of critical raw materials for the green and digital transition

As discussed earlier, raw materials are essential across all industrial sectors and along the entire supply chain. However, a subset of them is particularly crucial for the development of strategic sectors such as renewable energy, electric mobility, and digital technologies, making them key enablers of the twin green and digital transitions promoted by the *European Green Deal*. Critical raw materials such as lithium, cobalt, nickel, rare earths, and platinum are indispensable for the production of wind turbines, solar photovoltaic panels, batteries, energy-efficient lighting, hydrogen production and storage systems, and all digital technologies. Projections, therefore, point to a steep increase in their demand in the coming decades. This raises the risk that, with the transition toward climate neutrality, the current dependence on fossil fuels could be replaced with a dependence on raw materials, many of which come from abroad and for which global competition is intensifying (EC, 2020b). In fact, strategic technologies for decarbonization and digitalization do not require oil, coal, or natural gas but rely on minerals, metals, and other non-energy raw materials extracted from mines. Ensuring secure and affordable access to these critical raw materials is therefore becoming central to Europe's energy and digital security in the 21st century.

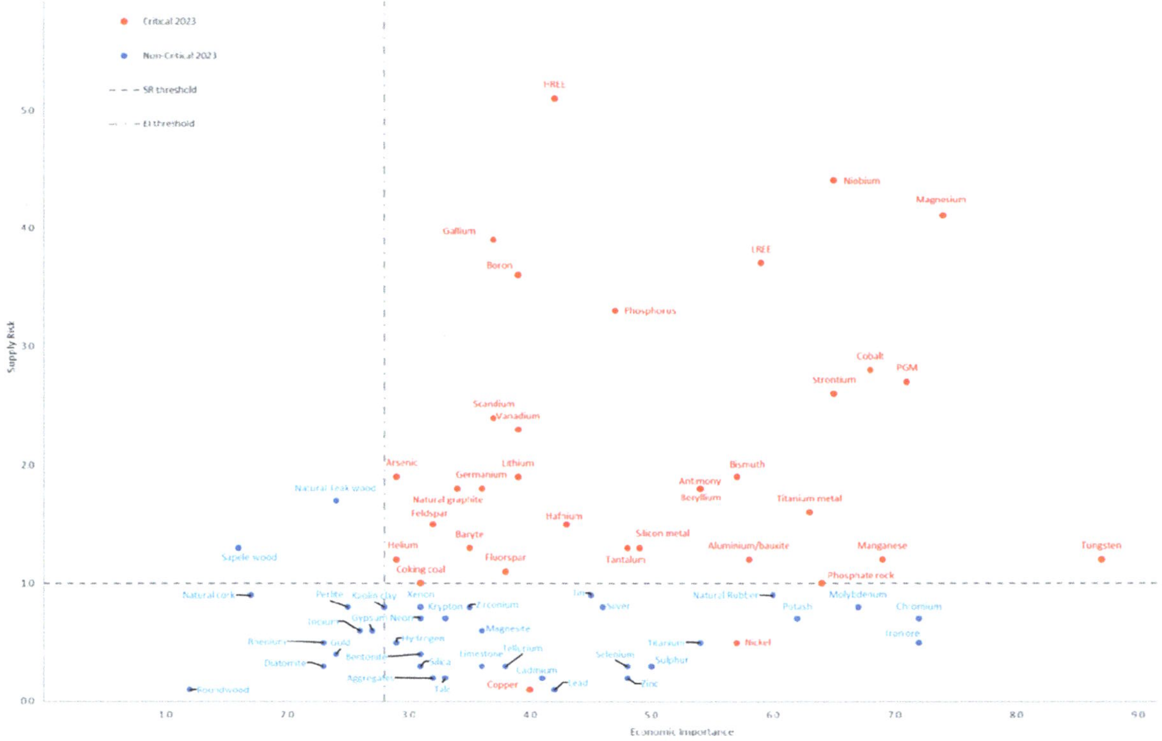

Figure 2.6 Critical raw materials (CRMs, in red) identified by the European Commission based on their economic importance (EI) and supply risk (SR). Copper and nickel are also reported in red as they are considered strategic raw materials (SRMs).

Source: EC, 2023a. © European Union, 2018. Reproduction is authorized provided the source is acknowledged.

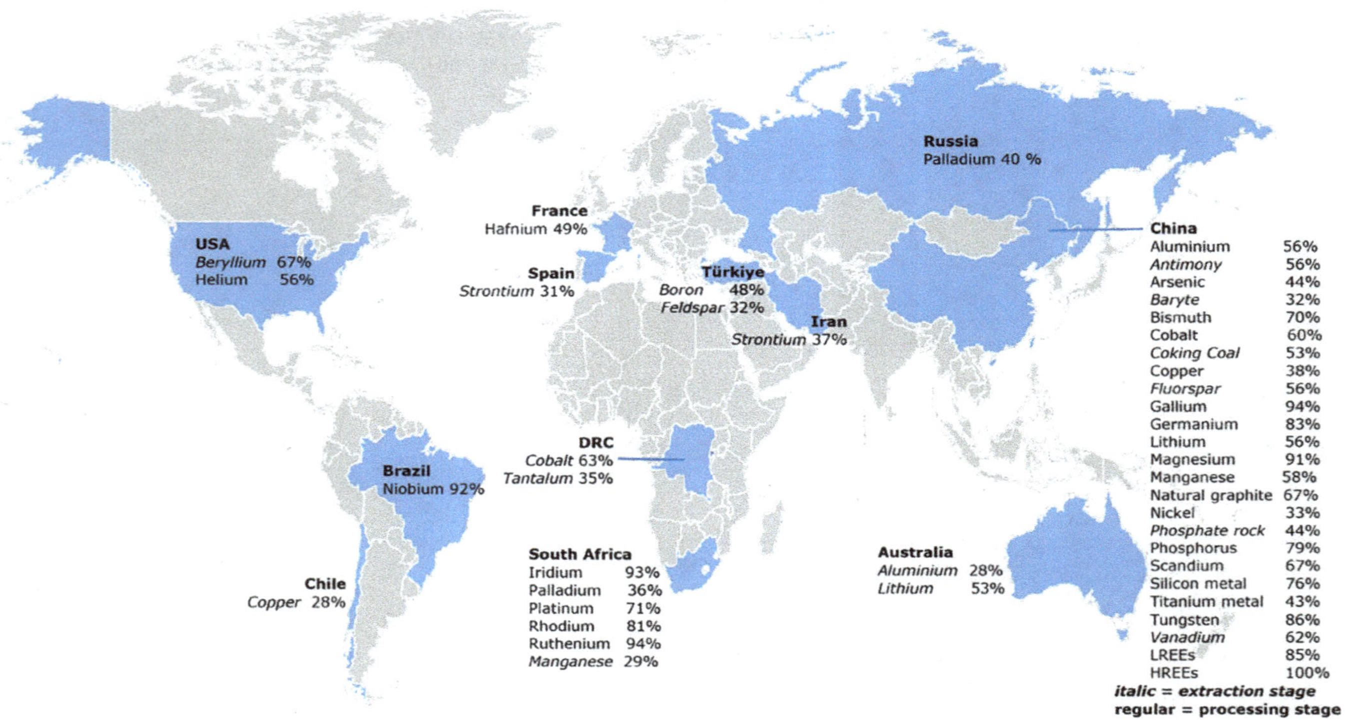

Figure 2.7 Map of the main global suppliers for each CRM.

Source: EC, 2023a. © European Union, 2018. Reproduction is authorized provided the source is acknowledged.

The COVID-19 crisis, along with recent geopolitical tensions and conflicts, has further exposed vulnerabilities of global supply chains. Whether through the temporary shutdown of mining and processing facilities or the interruption and restriction of key raw material supplies, these events have shown how quickly disruptions can affect the availability of critical materials.

Global demand for energy transition minerals is expected to quadruple by 2040 under a sustainable development scenario consistent with the Paris Agreement, i.e., with limiting global warming to 1.5°C and to increase sixfold by 2050 in a net-zero scenario (IEA, 2021). Electric vehicles and batteries will be the main drivers of this surge, followed by the large-scale expansion of electricity grids and renewable power technologies such as solar photovoltaic panels and wind turbines. In the latest report by the International Energy Agency (2025), the projections show that for batteries, demand for lithium is projected to grow fivefold by 2040, while the demand for graphite and nickel will double and for cobalt will increase 50%–60% by 2040. These materials are essential to ensure high performance and durability of battery technologies. For grid expansion, copper demand is expected to grow by 30% by 2040. Permanent magnets in wind turbines and electric vehicle motors will require REEs, with demand projected to increase by 2040. Solar panels rely heavily on copper, silicon, and silver, while fuel cells require platinum group metals.

What is most concerning is not only the magnitude of this expected growth in demand but also the geographic concentration of supply. Many key transition minerals are more geographically concentrated than fossil fuels such as oil or gas (IEA, 2021). As shown in Figure 2.8, the top three

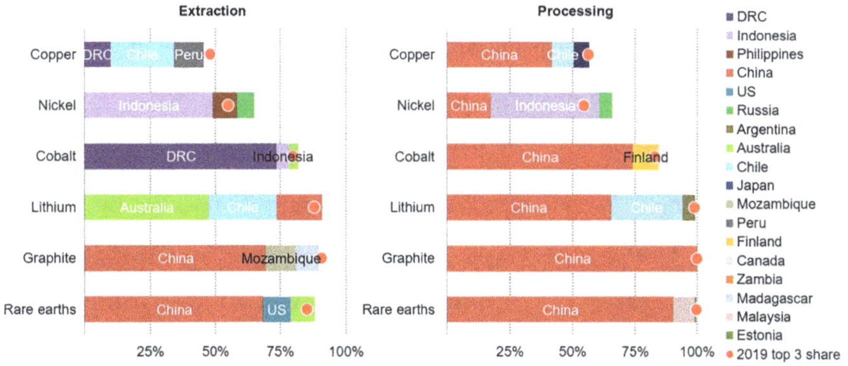

Figure 2.8 Share (%) of the top three producing countries in the production of selected minerals in 2022. The left-hand graph refers to raw material extraction, while the right-hand graph shows the subsequent processing stage.

Source: IEA, 2023. © IEA. Licensed under Creative Commons Attribution 4.0 International (CC BY 4.0).

producing countries account for between 50% and over 80% of global output for many CRMs, such as nickel, cobalt, lithium, graphite, and rare earths (IEA, 2023). In some cases, a single country dominates: the Democratic Republic of Congo provides around 75% of the world's cobalt, while China supplies roughly 60% of all REEs.

When moving downstream to processing, the concentration is even more striking: China holds a dominant role in many refining and processing stages of the most in-demand minerals, accounting for around 35% of copper refining, 65%–70% of lithium and cobalt processing, nearly 90% of rare earth element separation, and close to 100% of natural graphite refining (IEA, 2023). This dominant position is the result of massive and long-term investments in the entire value chain of SRMs to support domestic industries in key sectors such as batteries, wind turbines, and solar panel manufacturing.

The European Commission recently released an updated foresight study that explores the role of critical raw materials in 15 key technologies across five strategic sectors up to 2050 (Carrara et al., 2023), building on the previous foresight report (EC, 2020c). The results are summarized in Figure 2.9, which maps raw material flows distinguishing between CRMs and SRMs. The 15 technologies assessed include batteries, fuel cells, electrolyzers, wind turbines, traction motors, solar photovoltaic panels, heat pumps, hydrogen direct reduced iron and electric arc furnaces, data transmission networks, data storage and servers, smartphones, tablets and laptops, additive manufacturing, robotics, drones, and space launchers and satellites. The five strategic sectors considered are renewable energy, e-mobility, energy-intensive industry, ICT, and defense & aerospace. As shown in Figure 2.9, the renewable energy sector accounts for the largest share of SRM demand.

From the analysis of Figure 2.9, several aspects emerge as particularly relevant in the context of sustainable development and the twin green and digital transitions.

- REEs are among the raw materials with the highest SR, being essential for the production of permanent magnets. The strongest permanent magnets, NdFeB types, typically contain 27%–32% by weight of REEs (mainly neodymium and praseodymium, with additions of dysprosium, gadolinium, and terbium) (Zhang et al., 2020). Their size and lifetime vary widely, from a few grams with a 2- to 3-year lifespan in small electronics to several tons lasting decades in wind turbines (Yang et al., 2017). Permanent magnets are indispensable for clean technologies such as wind generators and electric motors, as well as for digital devices. China dominates the supply chain, covering almost the entire EU demand for REEs and controlling over 90% of the refining and alloy-making stages.

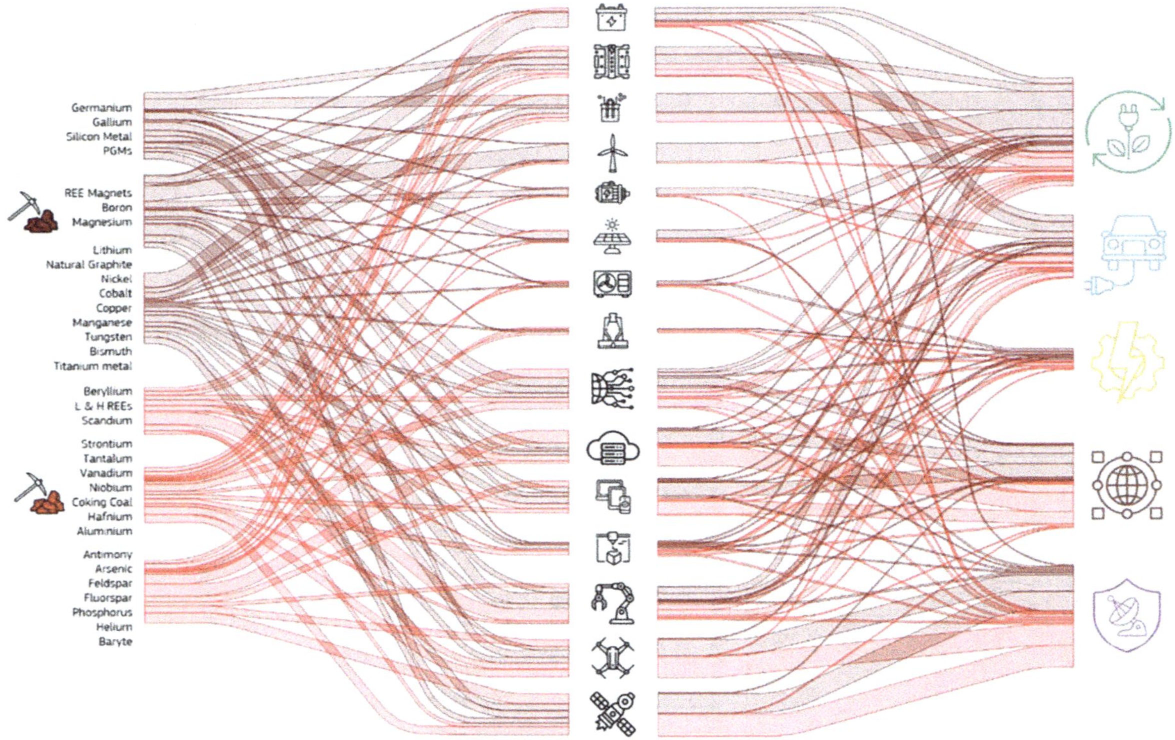

Figure 2.9 Flows of raw materials subdivided in SRMs and CRMs, related to fifteen key technologies and five strategic sectors.

Source: EC, JRC, Carrara et al., 2023, Reproduced under CC BY 4.0.

- Lithium, cobalt, nickel, and graphite are the key inputs for lithium-ion batteries, which power electric mobility, wind turbines, solar panels, and digital devices. A single EV battery typically requires around 50 kg of nickel, 45 kg of lithium, and 7 kg of cobalt. Europe currently relies on imports for about three-quarters of these resources, mainly from China, Africa, and Latin America, while its own production accounts for only about 1% (EC, 2020c). Lithium, in particular, is of increasing concern due to its unique properties, being the lightest metal, characterized by the highest electrochemical potential and energy density (Bae and Kim, 2021). Projections suggest that in a Paris Agreement compatible scenario, 90% of global lithium demand in 2040 will come from the energy sector (IEA, 2021).
- Platinum group metals (PGMs), especially platinum itself, are critical for fuel cell catalysts. South Africa is the largest producer, followed by Russia and Zimbabwe (EC, 2023a). In addition to the risks related to mining, Europe faces the challenge that only 1% of downstream assembly occurs within its borders.
- Base metals such as copper, aluminum, magnesium, and iron, together with alloying elements like tungsten, vanadium, manganese, and chromium, are needed across nearly all technologies mapped in the foresight study (EC, 2023a). Demand is also growing for specialized elements used in superalloys, including niobium, scandium, hafnium, and zirconium, each characterized by highly concentrated supply basis. Copper, in particular, is crucial for decarbonization given its unrivaled conductivity and use in renewables, electric vehicles and, most importantly, power grids. Its demand is projected to grow by 20%–40% by 2040, depending on the scenario (Copper factsheet, SCRREEN2, 2023).
- Digital technologies depend on a wide spectrum of the periodic table, including copper, gallium, germanium, gold, indium, PGMs, REEs, and tantalum. Europe's supply is mainly covered by China (41%) and Africa (30%).

These figures raise the critical question whether resource availability will keep pace with the rapidly increasing demand driven by the twin transitions. Projections for copper, lithium, and cobalt already suggest that new supply sources must be developed within the next decades to stay on track with a sustainable development pathway (IEA, 2025). This mismatch between climate ambitions and raw material availability underlines the urgency of concrete policy action and investment in the sector. Another key challenge is the declining ore grades of many deposits, as richer portions are mined first. For example, the average copper content of ores extracted in Chile has dropped by about 30% over the last 15 years (IEA, 2021). Lower grades imply greater energy use, higher costs, and increased greenhouse gas emissions and waste.

2.3.3 European Union action plan on critical raw materials

Based on the considerations discussed in Sections 2.3.1 and 2.3.2, Europe faces an urgent need to complement primary raw material extraction with alternative supplies strategies, for both environmental and economic reasons. In response to these challenges, the European Commission launched the European Raw Materials Alliance (ERMA) on 3 September 2020. ERMA is an open, multistakeholder network that brings together public and private actors from across the entire raw materials value chain.

The alliance aims to secure access to the mineral resources required to support the European Union's green and digital transition, while strengthening Europe's industrial base and reducing strategic dependencies on external suppliers, particularly China and other non-EU countries. Furthermore, the European Commission has developed the "Action Plan on Critical Raw Materials" (EC, 2020b), which not only established a systematic framework for identifying and monitoring CRMs over time but also defined a set of strategic intervention priorities:

- Strengthening Europe's domestic supply capacity by increasing extraction and processing activities within the EU and thereby reducing reliance on imports. This includes reassessing the potential reopening of closed mines and exploiting mining residues and tailings through advanced processing technologies.
- Ensuring secure and sustainable access to CRMs through diversification of supply from third countries, while promoting responsible mining practices and reducing distortions in international trade.
- Developing substitutes for critical raw materials wherever possible, using alternative materials that can deliver equivalent technical performance.
- Lowering overall primary raw material demand by accelerating the transition toward more efficient and circular resource use. In this context, sustainable product design, waste prevention, reuse, and recycling play a central role.

With regard to the last point, recycling represents a particularly strategic opportunity for Europe. Given the continent's limited primary mineral reserves, secondary raw materials offer not only economic advantages but also significant environmental benefits. They conserve non-renewable resources, reduce the environmental footprint of primary extraction, and contribute directly to climate-neutrality objectives. While recycling systems for widely used metals such as copper and aluminum are already relatively mature, recycling technologies for many materials that are critical to the twin transition, including lithium and REEs, remain at an early stage of development. Investing in research, development and innovation in the recycling sector is therefore essential, especially in light of two converging trends:

1. The forthcoming wave of emerging waste streams generated by new technologies, such as end-of-life batteries and wind turbines;
2. The fact that many technologically valuable metals occur in higher concentrations in complex waste streams than in primary mineral ores. As a result, large volumes of end-of-life products will become an increasingly important source of SRMs, including copper, lithium, nickel, cobalt, rare earths, and many others.

In this framework, the actual contribution of recycling to the supply of many raw materials, including critical ones, is still very low, as shown in Figure 2.10 by the so-called *end-of-life recycling input rate* (EOL-RIR), which measures the share of secondary raw materials in total material demand (EC, 2021). Circular use rates differ substantially from one metal to another. Lead exhibits the highest value, around 75%, and is the only element exceeding the 50% threshold. About 15 elements display intermediate circularity levels between 25% and 50%, including several critical raw materials for the EU such as tungsten (42%), europium (38%), yttrium (31%), palladium and rhodium (both 28%), and platinum (25%). A further group of 16 elements falls in the 10%–25% range, comprising widely used

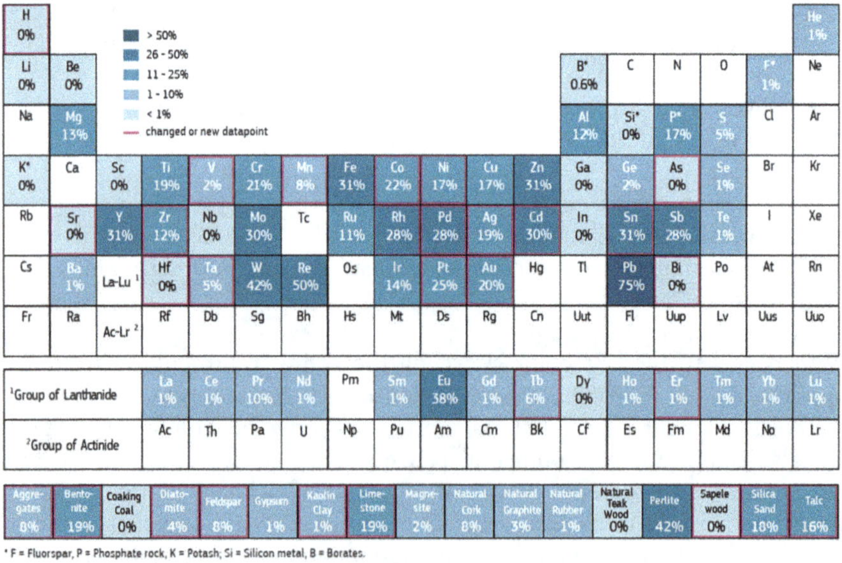

Figure 2.10 *End-of-life recycling input rate* (EOL-RIR) of different chemical elements/metals in Europe.

Source: EC, 2021. © European Union, 2018. Reproduction is authorized-provided the source is acknowledged.

metals such as aluminum, chromium, copper, magnesium, and nickel, as well as specialty and precious elements like iridium, ruthenium, and silver. Most alarming, however, is the fact that for many CRMs, especially REEs, the contribution of recycled material to overall supply remains negligible, often close to 1% or even lower.

The structural reasons behind these low circularity levels are multiple. In many cases, recycling is technically demanding and economically unattractive, since the critical elements are embedded in very small quantities within complex products containing many different materials. Under such conditions, primary extraction often remains the simpler and cheaper option. Furthermore, the availability of end-of-life materials is still limited for many CRMs because their large-scale deployment is relatively recent and a substantial share of products remains in active use. Lithium provides a clear illustration: despite its rapidly expanding demand, its current recycling rate is still very low.

These observations further reinforce the central role of circular economy strategies for technological waste streams. In this context, several key stages of the value chain require coordinated action. First, product design and manufacturing must prioritize durability, repairability, and recyclability. Second, collection systems are decisive for closing material loops, whereas currently represent one of the weakest points of the chain; the development of digital tools, product passport and traceability mechanisms is therefore essential. Third, recycling technologies must become progressively more efficient and consistent, in order to deliver high-quality secondary raw materials and significantly increase the circular use rates of CRMs.

2.3.4 The EU Critical Raw Materials Act

In response to the growing concern regarding the security of CRMs supply for the dual green and digital transitions, the EU adopted the Critical Raw Materials Act (CRMA) in 2023, which entered into force in 2024 (EC, 2023b, 2024). The Act establishes a comprehensive regulatory framework aimed at ensuring the EU's long-term access to a secure, resilient, sustainable, and diversified supply of critical and strategic raw materials, which are indispensable for the green and digital transitions as well as for defense and space technologies.

The CRMA introduces quantitative benchmarks to be achieved by 2030 along the entire value chain of SRMs:

- At least 10% of EU's annual consumption from domestic extraction,
- At least 40% of EU's annual consumption from processing,
- At least 15% of EU's annual consumption from recycling, and
- No more than 65% of EU's annual consumption of any strategic raw material sourced from a single third country at any relevant stage of processing.

To achieve these objectives, the Regulation streamlines permitting procedures for SRMs projects, introduces a fast-track regime for strategic projects, and strengthens financial and administrative support for investments in extraction, processing, and recycling capacities within the EU. Member States are required to develop national exploration programs and to enhance geological knowledge of domestic resources.

A central pillar of the Act is the reinforcement of circularity and sustainability. The CRMA mandates improved collection and recycling of waste streams rich in CRMs, promotes recovery from current and historical mining waste, and introduces specific requirements for the recyclability and recycled content of permanent magnets. These measures are designed to increase the share of secondary critical raw materials in EU manufacturing while mitigating environmental and social impacts.

To support coordination and implementation, the Act establishes a European Critical Raw Materials Board, bringing together the Commission and Member States to oversee strategic projects, monitor supply risks, coordinate strategic stocks, and foster international partnerships.

Overall, the CRMA represents a structural shift in EU industrial and resource policy, embedding raw material security at the core of Europe's climate, industrial, and geopolitical strategy.

2.4 WASTE MANAGEMENT AND RECYCLING

2.4.1 Global and European waste generation

2.4.1.1 Global waste generation

Data on global waste generation across all sectors, including industrial activities, construction and demolition, mining and other material streams, are limited and fragmented at the international level. Recent assessments estimate that in 2017 the amount of solid waste generated worldwide reached approximately 19.8 billion tons, with projections indicating a dramatic increase to about 46 billion tons by 2050 (Maalouf and Mavropoulos, 2023).

According to the most recent Global Waste Management Outlook 2024 (UNEP, 2024b), annual municipal solid waste (MSW) generation, i.e., waste generated by households, commercial activities, and offices, has reached about 2.1 billion tons globally and is projected to increase to approximately 3.8 billion tons by 2050 under a business-as-usual scenario. This increase is driven by a combination of population growth, urban expansion, economic development and the diffusion of consumption-intensive lifestyles. These trends underline the urgency of transforming production and consumption systems in order to achieve the targets of SDG 12 and broader sustainability goals.

The average daily per-capita generation is 0.74 kg, with extremely wide regional variations ranging from 0.11 to 4.54 kg per person per day (Kaza et al., 2018). Waste generation is strongly correlated with income levels and urbanization patterns: high-income countries account for about 34% of global municipal waste, despite accounting for only 16% of the world's population. Projections further indicate that the most rapid growth in waste generation will occur in low- and middle-income regions, particularly Sub-Saharan Africa, South Asia, and the Middle East and North Africa, where increases exceeding 40% are expected by 2050. In these regions, more than half of the waste is currently disposed of in uncontrolled dumpsites, posing severe environmental, health and socio-economic risks. These dynamics make it clear that the transition toward circular economy models and resource-efficient systems must be pursued at the global scale, beyond national boundaries, as the environmental consequences of waste generation ultimately affect the planet as a whole.

2.4.1.2 European waste generation

In 2022, total waste generation amounted to approximately 2.23 billion tons (Eurostat, 2024). This quantity is distributed among different economic activities and households, as illustrated in Figure 2.11. If construction and

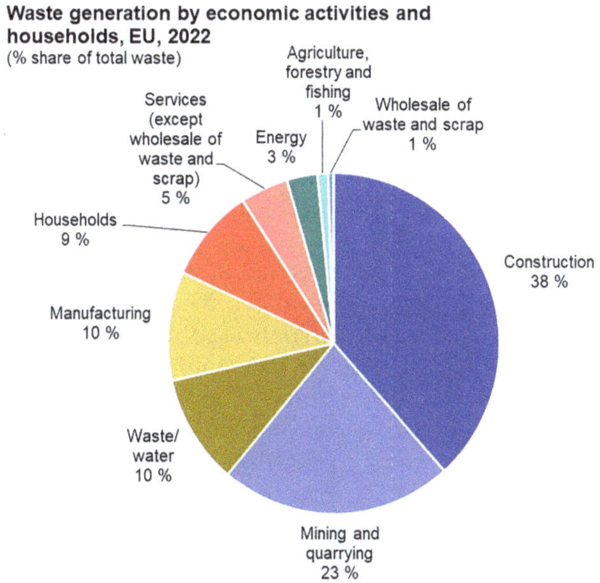

Figure 2.11 - Waste generation by economic activities and households, EU, 2022 (% share of total waste).

Source: Eurostat, online data code: env_wasgen. Reuse is authorized provided the source is acknowledged.

demolition waste (CDW) and mining and quarrying waste are excluded, the remaining total equals about 812 million tons, highlighting how strongly these two sectors dominate overall waste production. In fact, from the graph in Figure 2.11 it can be noticed as the construction sector is by far the largest contributor, accounting for 38% of total waste generation, followed by extractive industries, which contribute 23%. Together these sectors alone are responsible for more than half of all waste produced in Europe.

CDW represent a valuable source of secondary raw materials that can substitute for primary raw materials extracted from non-renewable resources. It is necessary to invest in scientific research, technological development, and innovation in this field to increase recycling rates and to produce high-quality secondary raw materials comparable to virgin materials, while removing contaminants (such as asbestos) that may hinder safe recycling (Serranti and Bonifazi, 2020). Concerning extractive activities, in addition to consuming non-renewable resources and generating significant environmental pressures, they also produce very large volumes of waste that must be properly managed. This underlines the need to improve the efficiency of primary production processes and to develop innovative solutions for the valorization of both newly generated waste and historical mining residues. Household waste represents 9% of total waste generation, corresponding to roughly 200 million tons. The remaining fractions are mainly associated with the manufacturing sector (10%), waste and water services (10%), and other economic activities, mainly other services (5%) and energy sector (3%).

2.4.2 The waste management hierarchy

The European Union's approach to waste management is structured around a hierarchical framework (Figure 2.12) that establishes clear priorities among the available waste management and treatment options. This hierarchy ranks the different actions according to their environmental performance and overall sustainability.

At the top of this hierarchy lies prevention, which includes all measures aimed at avoiding waste generation in the first place. Typical examples involve reducing the quantity of materials used in products, such as lightweighting plastic and glass bottles or aluminum cans, extending product lifetimes, and promoting reuse. The next level is preparation for reuse, which covers activities such as checking, cleaning, and repairing products or components that have become waste, so that they can be put back into use. This is followed by recycling, where waste materials are processed through industrial treatments to produce secondary raw materials such as metals, plastics, paper, and glass. When recycling (i.e., material recovery) is not feasible, energy recovery may be applied, allowing the generation of energy in waste-to-energy plants or through biogas production. At the bottom of the hierarchy, and therefore the least preferred option, is disposal in landfills, which

Figure 2.12 Waste management hierarchy pyramid, from the most sustainable option (prevention) to the least sustainable one (disposal).

Source: European Commission https://environment.ec.europa.eu/topics/waste-and-recycling/waste-framework-directive_en. Reproduction is authorized provided the source is acknowledged.

should be limited as much as possible and reserved only for waste streams that cannot be recovered by the previous methods.

In line with this hierarchy, European legislation sets specific targets for increasing recycling rates across key waste streams, including electrical and electronic equipment, end-of-life vehicles, batteries, CDW, municipal waste, and packaging waste.

2.4.3 Treatment methods for municipal solid waste, recycling rates, and EU targets

Among the different waste management options, particular attention must be given to MSW, as this is the waste stream most directly connected to everyday life and to individual responsibility for waste generation. Although MSW represents only about 9% of the total waste generated in Europe (Eurostat, 2024), it requires special attention because of its higher heterogeneity and the complexity of its management. In 2023, approximately 219 million tons of municipal waste were treated in Europe, representing a slight decrease compared to previous years. Of this total amount, about 77% underwent recovery processes with 30% material recycling, 27% energy recovery or incineration (the latter less than 1%), and 20% to composting and anaerobic digestion, while the remaining 23% was disposed of through landfilling.

In recent years, a positive trend has emerged, characterized by a gradual reduction in the share of waste sent to disposal and a corresponding increase in recovery-oriented treatment options.

Table 2.6 summarizes the main MSW treatment pathways in the European Union for each Member State for the year 2023 (ISPRA, 2025). These include material recovery operations, namely recycling of dry fraction and composting-anaerobic digestion, energy recovery, incineration, as well as disposal operations, such as landfilling and other disposal routes.

As shown in Table 2.6, MSW treatment practices vary considerably across European countries. Some Member States display highly advanced waste management systems, with recovery rates approaching 100%, including Belgium, Finland, Estonia, Germany, the Netherlands, Sweden, Austria, and Denmark. In contrast, other countries still rely heavily on landfilling, with disposal shares exceeding 70% in Greece, Romania, Malta, and Cyprus. The high share of waste sent to landfill in several countries is particularly concerning, as the European Union has established a binding target under the Circular Economy Package, which entered into force in July 2018, requiring the landfilling of municipal waste to fall below 10% of total generation by 2035.

A noteworthy pattern emerges when comparing disposal and recovery options: countries with the lowest landfill rates tend to exhibit significantly higher shares of energy recovery, reflecting the widespread availability of waste-to-energy facilities.

Regarding material recovery, which includes both recycling and composting or anaerobic digestion, several Member States already exhibit advanced circular management pathways, with combined recovery rates exceeding 55% of total municipal waste, such as Slovenia, Germany, Austria, Italy, the Netherlands, Luxembourg, and Belgium.

This performance is particularly relevant in light of the progressive recycling targets introduced by the EU Circular Economy Package, which require that at least 55% of municipal waste be recycled by 2025, 60% by 2030, and 65% by 2035 (Directive EU 2018/851).

Overall, these trends clearly indicate that, despite recent progress, substantial efforts are still required across most Member States to achieve the waste management objectives set by European policy.

2.4.4 Packaging waste and EU recycling targets

A particularly relevant and distinct category of municipal solid waste is represented by packaging waste, for which the European Union has established a dedicated regulatory framework with specific recycling targets. According to Directive (EU) 2018/852 on packaging and packaging waste, packaging is defined as: “all products made of any materials of any nature to be used for the containment, protection, handling, delivery, and presentation

Table 2.6 Percentage distribution of municipal waste management options in EU27 countries in 2023.

Country	*Disposal - landfill and other (D1–D7, D12) %*	*Energy recovery (R1) & incineration (D10) %*	*Composting and anaerobic digestion %*	*Recycling of dry fraction %*
Belgium	0	44	21	34
Finland	0	55	16	29
Estonia	1	47	11	31
Germany	1	30	23	46
Netherlands	1	40	31	28
Sweden	1	59	19	20
Austria	2	35	21	41
Denmark	2	51	19	27
Luxembourg	3	41	25	32
Lithuania	8	41	24	25
Slovenia	11	16	18	54
Italy	18	23	27	33
EU27	**23**	**27**	**20**	**30**
France	23	35	20	22
Poland	39	26	15	20
Slovakia	40	8	18	33
Czech Republic	42	14	15	28
Spain	48	11	21	21
Portugal	54	17	17	12
Hungary	54	12	14	19
Croatia	58	2	7	33
Bulgaria	65	5	4	21
Cyprus	77	3	4	16
Malta	79	3	0	18
Romania	80	6	5	8
Greece	81	2	1	16

Source: ISPRA (2025). The table reports the shares of landfilling and other disposal operations (D1–D7, D12), energy recovery and incineration (R1, D10), composting and anaerobic digestion, and recycling of the dry fraction. Countries are ordered by increasing share of landfilled municipal waste in 2023.

of goods, from raw materials to processed goods, from the producer to the user or the consumer."

Packaging is characterized by the shortest life cycle among consumer products and therefore becomes waste extremely rapidly. For this reason, prevention and reduction of packaging use represent key priorities of European waste policy.

Packaging waste is classified into the following material categories: steel, aluminum, paper and cardboard, wood, plastic, and glass.

In 2023, 79.7 million tons of packaging waste were generated in the European Union, corresponding to 177.8 kg per inhabitant, marking a reduction of 8.7 kg per capita compared with 2022 (Eurostat, 2024).

Figure 2.13 illustrates the distribution of packaging waste by material type in the EU in 2023 (Eurostat, 2025). Paper and cardboard represent the largest share (40.4%), followed by plastic (19.8%), glass (18.8%), and wood (15.8%), while metals and other materials account for smaller fractions. This distribution reflects both consumption patterns and the widespread use of paper-based packaging in commercial and logistics activities, while the significant shares of plastic highlight the relevance of this stream for recycling policies.

In Directive (EU) 2018/852, the EU establishes highly ambitious recycling targets for packaging waste, defined both for total packaging waste and for each individual material fraction, with percentages that increase progressively from 2025 to 2030 (Table 2.7).

Figure 2.14 reports the recycling rates of packaging waste in 2023 across EU Member States, indicating the 2030 target of 70%. On average, the EU has almost reached the target of 70% set for 2030, with a recycling rate of 67.5%. However, the figure reveals substantial heterogeneity among countries: several Member States have already achieved the 70% threshold, including Belgium, the Netherlands, Italy, Czechia, Slovenia, Slovakia, and Spain, while others remain significantly below the target, particularly Romania, Hungary, Malta, and Greece, indicating that considerable efforts are still required to meet the forthcoming regulatory obligations. The

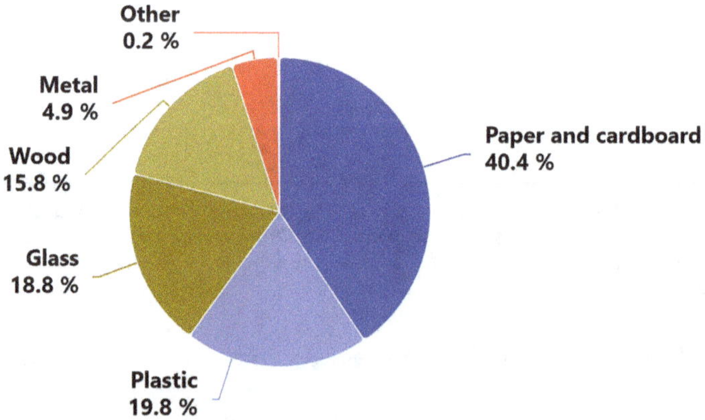

Figure 2.13 Packaging waste generated, by packaging material, EU, 2023.

Source: Eurostat, 2025, https://ec.europa.eu/eurostat/statistics-explained/index.php?title=Packaging_waste_statistics. Eurostat, online data code: env_waspac. Reuse is authorized provided the source is acknowledged.

Table 2.7 EU recycling targets for packaging waste by material and overall (Directive EU 2018/852)

	Paper	*Glass*	*Plastic*	*Aluminum*	*Steel*	*Wood*	*Overall packaging*
Target 2025 (%)	75	70	50	50	70	25	65
Target 2030 (%)	85	75	55	60	80	30	70

comparison between current performance and future targets underscores the urgency of strengthening collection systems, improving recycling technologies and expanding markets for secondary raw materials.

Altogether, packaging waste management constitutes a strategic lever for advancing the circular economy in Europe, not only because of the large quantities involved but also due to the high potential for recovering valuable secondary raw materials which are essential for reducing dependence on primary resource extraction and for supporting the EU's climate and resource efficiency objectives.

Plastic packaging represents the most critical fraction in the context of packaging recycling.

The term plastic encompasses a wide spectrum of chemically distinct polymers, each identified by a specific resin code introduced to support sorting and recycling operations (ASTM, 2014).

The primary challenge lies in the need to separate these polymers with high purity: when different polymer types are processed together, the resulting recycled material suffers a significant loss of mechanical and functional properties, leading to secondary raw materials of markedly lower quality than virgin plastics (Serranti and Bonifazi, 2018). This intrinsic complexity makes continuous investment in research and technological innovation essential to improve sorting accuracy and enhance the performance of polymer recycling processes.

Beyond technical challenges, plastic waste management constitutes one of the most pressing environmental issues of our time, particularly for packaging plastics, which are characterized by extremely short service lives, often shorter than 1 year. Recognizing this urgency, the European Union has devoted specific policy attention to the plastics value chain through the European Strategy for Plastics in a Circular Economy (COM, 2018). The strategy aims to increase plastic recycling rates, prevent the release of hazardous substances into the environment, and promote the development and market uptake of alternative materials, such as biopolymers, capable of providing comparable functionality with reduced environmental impact.

The strategy establishes that by 2030 all plastic packaging placed on the EU market must be reusable or recyclable. In addition, to curb plastic waste generation and extend product lifetimes, the EU subsequently adopted Directive (EU) 2019/904 on single-use plastics (SUP Directive). This directive

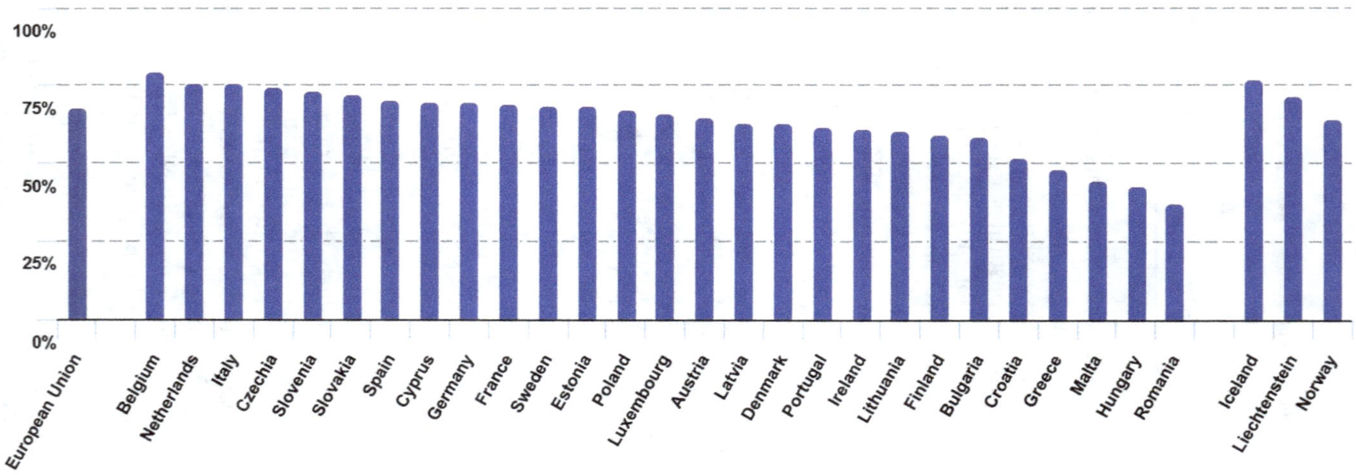

Target 2030: 70%

Note: sorted by recycling rate. EU: Eurostat estimate. Croatia, Czechia, Finland, Hungary, Poland, Slovakia: provisional. Cyprus: estimate. Bulgaria, Cyprus, Romania: 2022 data instead of 2023. Iceland, Slovakia: definition differs.

Figure 2.14 Recycling rate of packaging waste, 2023 (%).

Source: Eurostat, 2025, https://ec.europa.eu/eurostat/statistics-explained/index.php?title=Packaging_waste_statistics. Eurostat, online data code: env_waspacr. Reuse is authorized provided the source is acknowledged.

introduces a comprehensive set of measures, including bans on selected single-use plastic products, collection targets for plastic beverage bottles (90% by 2029), mandatory incorporation of at least 30% recycled plastic in beverage bottles by 2030, and specific design requirements such as tethered caps.

2.4.5 The recycling process

The recycling of waste materials involves a sequence of physical/mechanical operations designed to recover individual material fractions from heterogeneous waste streams.

The overarching goal of this process is to separate the different components of waste as efficiently as possible in order to produce high-quality secondary raw materials that can effectively replace primary resources in manufacturing processes.

The material recycling chain typically consists of several interconnected stages. The most common operations include collection, disassembly, manual sorting, size reduction (comminution), screening, and a variety of material separation techniques. These operations are generally carried out in a specific sequence, although some steps may be repeated or omitted depending on the characteristics of the input waste stream, such as its composition, particle size distribution, shape, and degree of contamination (Bonifazi and Serranti, 2012).

The final destination and market value of the recovered secondary materials depend critically on their properties, in particular on their purity, which must comply with the technical requirements of downstream industrial applications. For this reason, process configurations are typically designed according to the desired output quality rather than following a rigid standard scheme.

Figure 2.15 schematically illustrates the main stages of a typical material recycling chain. Not all operations are necessarily present in every plant as the configuration of the process depends on the nature of the waste being treated. For instance, size reduction can be applied either before or after preliminary separation in order to optimize the overall efficiency of the recycling process.

2.4.5.1 *Collection*

All waste recovery activities start with the collection phase, which is organized through different systems depending on the origin of the waste stream, for example, whether it comes from households or industrial activities.

Figure 2.15 Schematic sequence of the main operations involved in the materials recycling process.

During separate collection, waste is sorted into one or more material fractions. Collection can be mono-material, when a single material is collected independently (for example, glass in dedicated containers), or multi-material, when several material fractions are collected together (such as plastics and metals).

Once collected, the waste is transported to sorting facilities for further treatment. It is important to underline that separate collection is not an end in itself, but rather a key enabling step that facilitates the efficiency of the subsequent treatment processes.

2.4.5.2 Disassembly

This operation is necessary for complex waste, i.e., end-of-life products composed of multiple interconnected components. They include waste from electrical and electronic equipment (WEEE) and end-of-life vehicles (ELVs). A similar preliminary operation is also carried out for construction and demolition waste (CDW) through selective demolition. The objectives of disassembly include: (i) enhancing the recovery of components and materials, (ii) preparing waste for subsequent processing steps, (iii) reducing the volume of mixed residual waste, and (iv) removing hazardous materials when present.

Disassembly procedures should ideally be planned already at the product design stage according to the principles of *Design for Disassembly*, in order to facilitate repair and reuse where possible, and recycling when reuse is no longer feasible, fully in line with circular economy principles.

Disassembly lines may be manual or automated, and the development of advanced automated dismantling systems for complex products such as WEEE and ELVs is progressing rapidly alongside Industry 4.0 technologies and robotics.

2.4.5.3 Manual sorting

Manual sorting is usually carried out by operators who inspect the waste stream as it moves along conveyor belts, removing contaminants and bulky objects.

This operation is usually positioned at the beginning of the recycling process, but it may also appear at other stages of the plant, for example, downstream of separation units, where it is used to refine the purity of the sorted material by removing residual contaminants that were not completely eliminated by the automated separators, which are inherently subject to classification errors.

Manual sorting is physically demanding work, as operators spend long periods standing and performing rapid, repetitive movements, often under difficult environmental conditions (including exposure to dust, noise, odors, microorganisms, and potentially hazardous substances).

Although technological progress and the development of robotics are gradually leading to increased automation of sorting operations, manual sorting remains a widely used practice in current recycling plants.

2.4.5.4 Comminution

Comminution aims at reducing the size of waste materials in order to achieve volume reduction, promote homogenization, liberate constituent components (especially in composite materials), and prepare the material for subsequent mechanical or chemical treatment steps.

Size reduction is performed using machines equipped with dedicated tools that apply mechanical actions such as impact, compression, cutting, and abrasion. The choice of the comminution method depends primarily on the nature of the input material and on its initial size distribution.

The most widely used comminution units in recycling plants are hammermills, operating at high rotational speeds and mainly applying impact forces, and shredders, where size reduction is dominated by cutting actions. Hammermills are typically equipped with discharge grates that allow particles to exit only once they reach a size smaller than the grid openings.

Comminution may be carried out either before or after separation, depending on the process configuration and on the characteristics of the waste stream.

2.4.5.5 Screening

Screening is employed to classify materials according to particle size by means of screening surfaces. Oversized particles remain on the screen surface, while finer fractions pass through the apertures. This operation allows the removal of undesired size fractions, the preparation of materials with controlled size classes for downstream separation processes, and the elimination of very fine particles, such as glass fragments, soil, and stones, that would otherwise hinder subsequent separation stages.

Screening can be performed either in wet or dry conditions, with dry screening being the most common approach in recycling facilities. The main screening technologies include rotating drum screens (trommels), vibrating screens, and disc screens, each characterized by specific mechanical configurations and operating principles. Trommel screens consist of large rotating cylindrical drums equipped with perforated screening surfaces. As the drum rotates around a slightly inclined horizontal axis, the material is progressively transported while being separated according to size. Vibrating screens are flat screening units composed of one or more decks fitted with apertures of selected dimensions; vibration promotes the efficient size separation. Disc screens are formed by a series of parallel rotating shafts fitted with rounded discs of various shapes and spacing. The combined motion of the shafts conveys the material while separating it according to particle size.

2.4.5.6 Material separation

Material separation aims at producing streams of the highest possible purity by exploiting differences in physical and chemical properties among materials. Widely applied separation methods include magnetic, electrostatic, eddy current, gravimetric, aerodynamic, optical, and spectroscopic techniques.

Separation is considered direct when it relies on intrinsic material properties, such as magnetic susceptibility, electrical conductivity, or density, through the application of external fields. Table 2.8 summarizes some of the main direct separation methods commonly applied in recycling plants.

In contrast, indirect separation involves prior identification of the materials by sensors, followed by mechanical ejection, for example, by compressed air jets, at the end of the conveyor belt. Sensor-based sorting represents the most advanced separation methods, employing techniques such as X-ray analysis and near-infrared spectroscopy for material recognition.

Table 2.8 Main methods of direct separation

Operation	*Method*	*Description*
Gravimetric separation	Sink-float separators	When materials are introduced into a tank containing a fluid of a specific cut density, materials with lower density float while those with higher density sink (e.g., separation of polymers of different densities).
	Air classifiers	Waste is fed into a column in which there is an ascending airflow. Light fractions (e.g., paper, plastic film, etc.) are transported upwards while heavier ones fall downwards.
	Ballistic separators	An inclined screening surface consisting of a series of parallel paddles with orbital motion. Light and flat fractions (2D) are carried to the top, heavy and rolling ones (3D) fall down. Fine fractions pass through the meshes of the screening surface.
Magnetic separation	Magnetic drum, head-pulley, and overbelt separators	Ferrous metals are separated from other materials by means of magnets that attract them, diverting them from the waste stream moving on the conveyor belt.
Eddy current separation	Induced current separator (or Eddy current)	Eddy currents generated by a rotating magnetic field are used to divert non-ferrous metals (e.g., aluminum) from the waste stream transported on a conveyor belt.
Electrostatic separation	"Corona effect," induction or triboelectric electrostatic separators	Materials are separated according to different electrical conductivity (e.g., metals from non-metals, organic fraction from inert one or different polymers). The methods differ according to the charging mechanism employed.

Table 2.9 Main types of sensors utilized in recycling plants and their typical applications

Sensor type	*Typical application*
Electromagnetic sensor (metal detector) - EM	Sorting of metals from non-metals (inert, plastic, glass, etc.)
X-ray transmission - XRT	Sorting of aluminum from heavy metals
X-ray fluorescence - XRF	Sorting of single metals and alloys; lead glass and ceramic glass from glass
Visible spectroscopy - VIS	Sorting of polymers, glass, or metals by color
Near-infrared spectroscopy - NIR	Sorting of polymer types

A typical sensor-based sorting system consists of three main stages:

- Feeding: during this stage material dispersion is ensured, often through vibrating devices, to prevent particle overlapping.
- Detection and analysis: sensors identify the properties of individual objects through non-contact and non-destructive measurements. The type of information collected depends on the specific sensor technology employed. The sensor system is connected to a data-processing unit that classifies each object according to the detected characteristics. These processing systems must operate with extremely high computational speed in order to handle the large data volumes generated at the high conveyor belt velocities typically used in industrial plants (about 2–4 m/s).
- Extraction: based on the classification results, the data-processing unit sends control signals to a series of actuators, usually compressed air nozzles positioned at the end of the conveyor belt. These actuators are selectively activated to deflect individual objects and separate the different material classes into distinct product streams. Table 2.9 lists the main types of sensors and their corresponding applications.

Once the different material fractions have been separated, such as paper, glass, aluminum, plastics, and others, they can be reintroduced into industrial production chains as secondary raw materials.

The processing stages described so far are primarily based on physical principles and rely on mechanical treatment methods for material separation. However, when dealing with highly complex products, for instance, printed circuit boards from WEEE, mechanical recycling alone is insufficient to fully liberate the valuable and critical metals embedded in composite structures. In such cases, additional chemical treatment processes are required, such as hydrometallurgical approaches, which exploit reactions in aqueous solutions to selectively extract metals from complex material matrices.

REFERENCES

ASTM (2014). *ASTM D7611 Standard Practice for Coding Plastic Manufactured Articles for Resin Identification*. ASTM International, West Conshohocken, PA.

Bae, H., and Kim, Y. (2021). Technologies of lithium recycling from waste lithium ion batteries: A review. *Materials Advances*, 2, 3234.

Bonifazi, G., and Serranti, S. (2012). Recycling Technologies. In: Meyers, R. A. (eds.), *Encyclopedia of Sustainability Science and Technology*. Springer, New York. https://doi.org/10.1007/978-1-4419-0851-3_116

Carrara, S., Bobba, S., Blagoeva, D., Alves Dias, P., Cavalli, A., Georgitzikis, K., Grohol, M., Itul, A., Kuzov, T., Latunussa, C., Lyons, L., Malano, G., Maury, T., Prior Arce, Á., Somers, J., Telsnig, T., Veeh, C., Wittmer, D., Black, C., Pennington, D., and Christou, M. (2023). *Supply Chain Analysis and Material Demand Forecast in Strategic Technologies and Sectors in the EU – A Foresight Study*, Publications Office of the European Union, Luxembourg. https://doi.org/10.2760/386650, JRC132889

COM (2018). A European Strategy for Plastics in a Circular Economy, 2018, 28 final.

COM (2019). The European Green Deal, 640 final.

Cossu R. (2013). The urban mining concept. *Waste Management*, 33, 497–498.

Cossu R., and Williams I.D. (2015). Urban mining: Concepts, terminology, challenges. *Waste Management*, 45, 1–3.

Ellen MacArthur Foundation (2019). Circular economy systems diagram. Accessed 20 April 2026. https://www.ellenmacarthurfoundation.org/circular-economy-diagram

European Commission (2018a). *Report on Critical Raw Materials and the Circular Economy, 2018*. Publications Office of the European Union, Luxembourg.

European Commission (2018b). *EIP on Raw Materials, Raw Materials Scoreboard 2018*. Publications Office of the European Union, Luxembourg.

European Commission (EC) (2020a). *Study on the EU's List of Critical Raw Materials – Final Report*. Publications Office of the European Union, Luxembourg.

European Commission (EC) (2020b). *Critical Raw Materials Resilience: Charting a Path Towards Greater Security and Sustainability*. COM (2020) 474 final. European Commission, Brussels.

European Commission (EC) (2020c). *Critical Materials for Strategic Technologies and Sectors in the EU - A Foresight Study 2020*. Publications Office of the European Union, Luxembourg.

European Commission (EC) (2021). *EIP on Raw Materials, Raw Materials Scoreboard, 2021*. Publications Office of the European Union, Luxembourg.

European Commission (EC) (2023a). Study on the Critical Raw Materials for the EU 2023 – Final Report. Publications Office of the European Union, Luxembourg.

European Commission (EC) (2023b). *Proposal for a Regulation of the European Parliament and of the Council Establishing a Framework for Ensuring a Secure and Sustainable Supply of Critical Raw Materials (Critical Raw Materials Act)*. COM(2023) 160 final, 16 March 2023. European Commission, Brussels.

European Environment Agency (EEA) (2016). *Circular economy in Europe Developing the knowledge base*, Report n. 2/2016, 42 pp. ISSN 1977-8449.

European Union (2024). *Regulation (EU) 2024/1252 of the European Parliament and of the Council Establishing a Framework for Ensuring a Secure and Sustainable*

Supply of Critical Raw Materials. Official Journal of the European Union, L series, 2024.

Eurostat. (2024). *Waste statistics*. Statistics Explained. European Commission, Luxembourg. Available at: https://ec.europa.eu/eurostat/statistics-explained/index.php?title=Waste_statistics (accessed 3 July 2026).

International Energy Agency (IEA) (2021). *The Role of Critical Minerals in Clean Energy Transitions*, IEA, Paris https://www.iea.org/reports/the-role-of-critical-minerals-in-clean-energy-transitions

International Energy Agency (IEA) (2023). *Critical Minerals Market Review, 2023*. Paris.

International Energy Agency (IEA) (2025). *Global Critical Minerals Outlook, 2025*. Paris.

International Resource Panel (IRP) (2017). Assessing global resource use: A systems approach to resource efficiency and pollution reduction. In: Bringezu, S., Ramaswami, A., Schandl, H., O'Brien, M., Pelton, R., Acquatella, J., Ayuk, E., Chiu, A.S.F., Flanegin, R., Fry, J., Giljum, S., Hashimoto, S., Hellweg, S., Hosking, K., Hu, Y., Lenzen, M., Lieber, M., Lutter, S., Miatto, M., Singh Nagpure, A., Obersteiner, M., van Oers, L., Pfister, S., Pichler, P., Russell, A., Spini, L., Tanikawa, H., van der Voet, E., Weisz, H., West, W., Wijkman, A., Zhu, B., and Zivy, R. A. (eds.), *Report by the International Resource Panel*. United Nations Environment Programme, Nairobi, Kenya (100 pp.).

International Resource Panel (IRP) (2019). Global Resources Outlook 2019: Natural resources for the future we want. Oberle, B., Bringezu, S., Hatfield-Dodds, S., Hellweg, S., Schandl, H., Clement, J., and Cabernard, L., Che, N., Chen, D., Droz-Georget, H., Ekins, P., Fischer-Kowalski, M., Flörke, M., Frank, S., Froemelt, A., Geschke, A., Haupt, M., Havlik, P., Hüfner, R., Lenzen, M., Lieber, M., Liu, B., Lu, Y., Lutter, S., Mehr, J., Miatto, A., Newth, D., Oberschelp, C., Obersteiner, M., Pfister, S., Piccoli, E., Schaldach, R., Schüngel, J., Sonderegger, T., Sudheshwar, A., Tanikawa, H., van der Voet, E., Walker, C., West, J., Wang, Z., and Zhu, B.(eds.), *A Report of the International Resource Panel*. United Nations Environment Programme. Nairobi, Kenya, 426 pp.

Istituto Superiore per la Protezione e la Ricerca Ambientale (ISPRA) (2025). *Rapporto Rifiuti Urbani*. Edizione 2025, 419. ISPRA, Rome, Italy. Rapporti 419/2025. ISBN 978-88-448-1285-0.

Kaza, Silpa, Yao, L., Bhada-Tata, P., and Woerden, Frank Van. 2018. *What a Waste 2.0: A Global Snapshot of Solid Waste Management to 2050. Urban Development Series*. World Bank, Washington, DC. https://doi.org/10.1596/978-1-4648-1329-0

Maalouf, A., and Mavropoulos, A. (2023). Re-assessing global municipal solid waste generation. *Waste Management & Research*, 41(4), 936–947.

Nakamura, T., and Halada, K. (2015). *Urban Mining Systems*, Springer ed. Springer, Tokyo.

Organisation for Economic Co-operation and Development (OECD) (2015). *Material Resources, Productivity and the Environment, OECD Green Growth Studies*. OECD Publishing, Paris. https://doi.org/10.1787/9789264190504-en

SCRREEN2. (2023). *Copper Factsheet*. European Union Horizon 2020 Research and Innovation Programme. Available at: https://scrreen.eu/wp-content/uploads/2023/03/SCRREEN2_factsheets_COPPER.pdf (accessed 3 July 2026)

Serranti S., and Bonifazi, G.(2018). Techniques for separation of plastic waste. In: Pachego-Torgal, F. et al. (ed.), *Use of Recycled Plastics in Eco-efficient Concrete* (pp. 9–37). Woodhead Publishing (Elsevier), Cambridge.

Serranti, S., and Bonifazi, G. (2020). Detection and classification of asbestos and other contaminants in C&DW by advanced technologies. In: Pachego-Torgal, F. et al. (eds.), *Advances in Construction and Demolition Waste Recycling* (pp. 407–437). Woodhead Publishing (Elsevier), Cambridge.

United Nations (2015). *Transforming our world: the 2030 Agenda for Sustainable Development*. Resolution adopted by the General Assembly on 25 September 2015, A/RES/70/1. United Nations, New York. https://sdgs.un.org/2030agenda

United Nations Environment Programme (2024a). *Global Resources Outlook 2024: Bend the Trend – Pathways to a Liveable Planet as Resource use Spikes*. International Resource Panel. Nairobi.

United Nations Environment Programme (2024b). *Global Waste Management Outlook 2024: Beyond an Age of Waste – Turning Rubbish Into a Resource*. Nairobi. https://wedocs.unep.org/20.500.11822/44939

United Nations Environment Programme, International Resource Panel, Global Material Flows Database. (n.d.). https://www.resourcepanel.org/global-material-flows-database

Yang, Y., Walton, A., Sheridan, R., Güth, K., Gauß, R., Gutfleisch, O., Buchert, M., Steenari, B. M., Van Gerven, T., Jones, P. T., and Binnemans, K. (2017). REE recovery from end-of-life NdFeB permanent magnet scrap: A critical review. *Journal of Sustainable Metallurgy*, 3, 122–149.

Zhang, Y., Foquan, G., Su, Z., Liu, S., Anderson, C., and Jiang, T. (2020). Hydrometallurgical recovery of rare earth elements from NdFeB permanent magnet scraps: A review. *Metals*, 10, 841

Chapter 3

Sustainable mobility: Relevant SDGs: 11, 3, and 13

Luca Persia and Davide Shingo Usami

3.1 SUSTAINABILITY OF TRANSPORT SYSTEMS

Cities represent the heart of life and the driver of the economy. Approximately 70% of the Italian population lives in urban areas, now considered "metropolitan areas" due to the phenomenon of uncontrolled urban growth (urban sprawl).

In urban areas, the demand for mobility is increasingly pressing; almost all daily trips, whether for study, work, or leisure, take place within the cities themselves or in nearby areas. In recent years, mobility dynamics have gradually changed, becoming less predictable, as the percentage of regular trips related to work and study has decreased.

The result of the growth and changing nature of mobility demand has been a significant increase in the modal imbalance in favor of private vehicles, which has led to a reduction in the average speed of transport modes due to traffic congestion. It has been estimated that today a car equipped with the latest technologies travels in cities at roughly the same speed as carriages and horses did a century ago (Federico & Barbabella, 2006). Traffic congestion is considered by a large share of citizens to be the main cause of the deterioration of urban livability. It entails the occupation of urban and road space, with a consequent worsening of the quality and usability of the city, loss of time due to long and stressful periods spent in cars, air pollution that remains significant despite technological advances in vehicles and fuels, high noise levels, and a high number of traffic accidents.

Traffic congestion, noise, air pollution, social exclusion, and road safety concerns are all issues that stand in the way of more sustainable urban development.

From this perspective, transport systems are linked to many of the Sustainable Development Goals (SDGs) of the United Nations 2030 Agenda.

The objectives of the SDGs that are most directly linked to the sustainability of transport systems include the following:

DOI: 10.1201/9781003595700-4

- 3.6, *Halve the number of global deaths and injuries from road traffic accidents by 2020*. This target was not achieved by 2020; in fact, the global number of road traffic fatalities continues to increase. The United Nations has set a new target to halve road traffic deaths by 2030.
- 11.2, *Provide access to safe, affordable, accessible, and sustainable transport systems for all by 2030, improving road safety, in particular by expanding public transport, with special attention to the needs of people in vulnerable situations, women, children, persons with disabilities, and the elderly*. Here too, trends do not give reassuring signals: cities are expanding faster than their populations, leading to a reduction in population density and a weakening of the effectiveness and efficiency of public transportation.

Other goals whose achievement is affected by transport system performance certainly include Goal 13 *(Take urgent action to combat climate change and its impacts)* and Goal 3.9 *(Substantially reduce by 2030 the number of deaths and illnesses from hazardous chemicals and air, water, and soil pollution and contamination)*, among others.

The challenge for policymakers and local administrators is to balance the increasing demand for mobility and economic growth with the need to improve the sustainability of the system, ensuring an acceptable quality of life for all citizens. While it is clear that mobility based on private vehicles will continue to be an important component of transport policy planning, finding ways to encourage the use of more sustainable alternative systems (public transport, cycling, and walking) is the aim of an urban sustainable mobility policy.

3.1.1 Sustainability and COVID-19

In this context, the impact of the COVID-19 pandemic on mobility systems and their sustainability cannot be overlooked. This impact is still being strongly felt at the time of writing, and even in future scenarios that are hopefully optimistic regarding the disappearance of COVID-19, it will require structural changes to mobility systems.

The need to ensure adequate conditions for interpersonal distancing, in order to prevent new waves of virus transmission, will force capacity limits to be maintained on public transport vehicles. There is also uncertainty regarding users' behavioral response, never before having experienced such an emergency, who will presumably change some aspects of their travel patterns, seeking to avoid crowded situations whenever possible. All of this may lead, even in a future where COVID-19 has disappeared, to a real risk of modal shift toward private vehicles, both four- and two-wheeled, with negative consequences in terms of increased congestion, additional externalities such as pollution and road safety issues, and therefore reduced sustainability.

One of the greatest challenges in addressing the difficulties created by the COVID-19 emergency stems from the fact that the need to maintain adequate interpersonal distancing forces us to significantly modify the paradigm adopted so far to increase the sustainability of transport systems: concentrating residential and work settlements as much as possible, favoring high-capacity public transport systems with high user density.

This approach, however, cannot be abandoned, as doing so would trigger an explosion in private vehicle traffic and a drastic deterioration in the sustainability of our transport systems and in the quality of life, with extremely serious consequences for citizens' health in the coming decades.

Thus, the major transport challenge that our community faces in the immediate and near future is to reconcile these two aspects: interpersonal distancing and urban/social densification. These two seemingly irreconcilable aspects can find an appropriate coexistence in a new balance, involving significant changes in both transport demand and supply. This would allow the entire community to transform the severe crisis we are experiencing into an opportunity for growth and improvement in mobility and living conditions. For this to be achieved, an integrated package of strategies must be adopted, aimed at:

- Adapting demand to the reduction in supply resulting from the decreased unit capacity of vehicles
- Predict the evolution of demand following the reopening of activities in the coming months
- Constantly monitoring the evolution of the mobility system and identifying its critical issues
- Planning the supply conditions needed to meet demand safely
- Redesigning a future mobility model with a higher degree of sustainability and resilience

Therefore, improving the sustainability of mobility systems in post-COVID-19 scenarios requires a new approach to planning that, alongside traditional planning of supply systems (infrastructure, vehicles, services), also includes optimization of demand systems through structural changes in users' habits and behaviors.

It can also be said that the transition toward new mobility systems that are more sustainable and more resilient to external disruptions requires:

- The need to foster a culture of sustainability and physical well-being, based on transport modes with zero or very low environmental impact and positive effects on people's health—primarily walking and cycling. This would have significant impacts, especially in urban mobility contexts, and would require the necessary infrastructure to allow the most sustainable modes to move safely within the mobility system.

- The need to involve citizens in decision-making processes (crowdsourcing), drawing on the vast wealth of "collective intelligence" and encouraging the development of a necessary social awareness regarding the sustainability of transport systems.

3.1.2 The concept of sustainability and sustainable mobility

But how is sustainable mobility defined? Before analyzing this concept, it is necessary to define that of sustainable development, to which it is closely connected.

Sustainable development emerged in the 1980s in response to the need to reconcile economic growth with fair resource distribution and to better protect the environment and social well-being. The term was officially introduced in the 1987 document *Our Common Future*, also known as the Brundtland Report, by the World Commission on Environment and Development (WCED, 1987) and defined as "development that meets the needs of the present without compromising the ability of future generations to meet their own." Since then, and especially following the United Nations Conference on Environment and Development (UNCED) held in Rio de Janeiro in 1992, socioeconomic development and environmental protection have become recurring themes in territorial policy decisions. The most important outcome of the conference was Agenda 21, a broad and comprehensive action program for the 21st century, consisting of planning the actions to be undertaken at the global, national, and local levels by UN organizations, governments, and administrations in every area where human presence impacts the environment. Sustainable mobility can therefore be considered "the expression of sustainable development within the transport sector," as defined by the OECD in 1997 (OECD, 1997).

Sustainable mobility is a very important but at the same time vague concept. It is important because it highlights that the mobility of people and goods—while intertwined with the economic development of territories, the circulation of ideas, and cultural growth—also generates, as mentioned earlier, negative environmental impacts and energy resource consumption, thereby contributing to the deterioration of the environmental quality we leave to future generations. Despite the relevance of the issue due to its strong impacts on individuals' quality of life, transport sustainability is still a controversial subject. The ambiguity of the concept of sustainable transport can be explained by three factors (Gudmundsson, 2003):

- Environmental sustainability issue: the difficulty in identifying sustainable limits for environmental use;
- Economic allocation issue: the difficulty of defining the optimal contribution of each economic sector to solving sustainability-related problems;

- Social interconnection issue: the difficulty in independently assessing mobility sustainability due to the links between transport activities and other activities, individuals' location choices, and lifestyles.

Nevertheless, several institutions have attempted to provide a structured definition of sustainable mobility. According to the OECD (2002), sustainable mobility can be understood as a mobility system that does not endanger public health or ecosystems and reconciles the need for accessibility with:

- The use of renewable resources at a rate equal to or less than their regeneration rate;
- The use of non-renewable resources at a rate equal to or less than the rate of development of renewable alternatives.

The European Commission's Commission Expert Group on Transport and Environment has developed the following definition: *A sustainable mobility system*:

- Enables basic access and development needs of individuals, companies, and society to be met safely and in a manner consistent with human and ecosystem health, and promotes equity within and between generations;
- Is affordable, operates efficiently, offers a choice of transport modes, and supports a vibrant economy and local development;
- Limits emissions and waste to levels the planet can absorb, uses renewable resources within the regeneration capacity of the present generation, employs non-renewable resources at or below the rate of development of renewable alternatives, and minimizes land use and noise pollution.

Building on this definition, it is widely acknowledged that although the concept of sustainable mobility does not yet have a universally accepted boundary, its assessment consistently relies on the three foundational dimensions of sustainability (Jeon & Amekudzi, 2005). These environmental, social, and economic pillars (presented in Figure 3.1) offer a structured lens through which the long-term performance, equity implications, and resource impacts of mobility systems can be critically evaluated.

The concept of sustainable transport, therefore, tends to be considered operationally according to a three-dimensional approach that simultaneously takes into account the impacts of mobility on the environment (in terms of pollution and use of natural resources), on the economy (in terms of directly and indirectly generated costs), and on society (in terms of health impacts, safety, and equity).

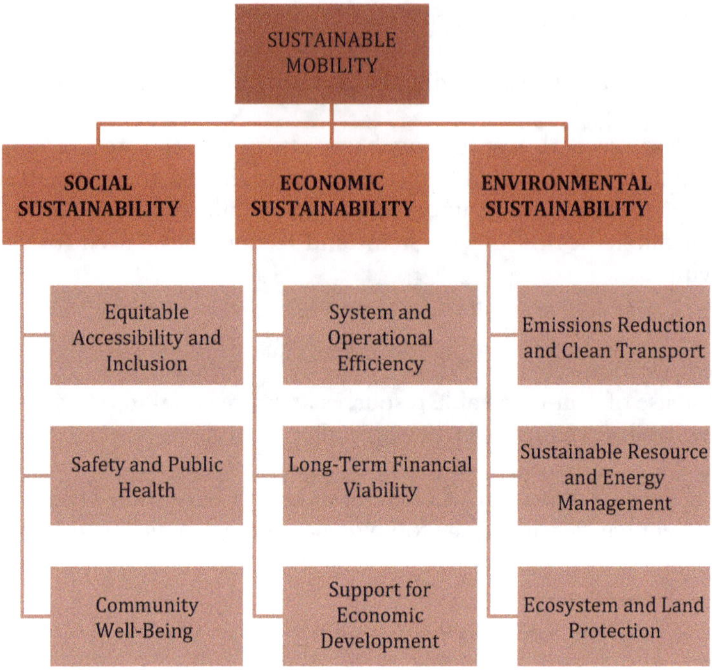

Figure 3.1 The three dimensions of sustainable mobility.

3.1.3 Indicators for measuring the sustainability of a transport system

Having recognized the importance of the concept of sustainability from economic, social, and environmental perspectives, the issue arises of how to measure it so that the concept becomes operational and can guide evaluation, public decision-making, and the implementation of more sustainable policies.

There is a broad consensus in research and planning on the use of indicators as assessment tools to describe and analyze the dimensions of sustainability. Indicators are an essential reporting tool for monitoring the performance of sustainable mobility policies, widely endorsed by numerous national, international, and European bodies as well as by scientific literature.

In general, it can be said that all phenomena related to sustainability are *quantifiable*, that is, expressible through numerical values: some are measurable phenomena (such as PM10 concentration), while others are characteristics that cannot be directly measured but can be quantified through appropriate weighting or classification methods (such as customer satisfaction among local public transport users).

In transport planning, the following categories are distinguished:

- *Input indicators*, which measure what has been done (e.g., the length of bus-priority lanes built);
- *Process indicators*, which describe how the transport system responds to the policies implemented (e.g., bus speed);
- *Output indicators*, which measure the outcome of a policy in terms of sustainability (e.g., pollutant concentrations or the number and severity of road accidents).

Indicators are therefore essential for assessing the level of sustainability of the transport system and, consequently, for implementing the most suitable policies for the context under consideration. However, the collection and analysis of data and indicators is a complex process, affected by several critical factors related to data accessibility, comparability between different contexts, quality, and cost. New technologies, such as the use of big data, can significantly improve this process.

3.1.4 Comparisons between cities: Italy, Europe, and worldwide

In 2009, the Transport and Logistics Research Centre of "Sapienza" University of Rome conducted a study within the framework of the General Mobility Plan and on behalf of the Ministry of Transport on improving the sustainability of Italian cities (CTL, 2009). The study aimed to identify the most effective policies for enhancing urban sustainability. A methodology was developed and applied to the 50 largest Italian cities through an analysis organized into the following mutually interacting steps: definition of indicators, clustering of cities into homogeneous groups, analysis of functional relationships between indicators, and identification of the most effective policies through benchmarking within the groups.

The *state indicators* (commuter mobility, vehicle fleet, overall urban mobility, and territorial variables), representing structural characteristics, were used to group cities with similar features (size and population density) so that they could be compared and the transferability conditions of successful policies and measures could be evaluated. The *sustainability indicators* (modal split of urban mobility, public transport demand, pollution, accident rates) were used to assess the current situation in each city, seeking to identify functional relationships between these and the *intervention indicators* (public transport supply, private transport supply, management measures, street space reallocation, and regulatory measures) in order to predict the outcomes of different types of policies.

The general analysis revealed that environmental sustainability, in terms of pollutant emissions, tends to improve in cities with higher public transport demand. It is also closely correlated with the intensity of public transport service and parking availability (high availability of parking worsens sustainability conditions). This result confirms the assumption that public transport is a key factor in improving the sustainability of urban mobility systems: compared to private vehicles, it is more efficient, occupies less road space, reduces the risk of accidents, lowers energy consumption, and provides mobility to those who, for various reasons, cannot use a private vehicle. Therefore, an increase in the share of users choosing this mode along with walking and cycling leads to improvements in sustainability levels.

Subsequently, the same methodology was applied to a sample of 45 European cities.[1] The analyzed data were taken from the "Mobility in Cities Database" project, conducted by UITP (the International Association of Public Transport). In this case, the relationship between land-use indicators, public transport indicators, private transport indicators, infrastructure provision, and transport system performance was examined to identify and analyze the variables associated with urban transport that significantly influence sustainability characteristics. The study showed a substantial diversity in urban mobility patterns across Europe, particularly between Eastern European cities and more economically developed Western cities. From the standpoint of urban sustainability, using variables such as average car occupancy rate, annual public transport passenger kilometers per capita, investment expenditure as a percentage of Gross Domestic Product (GDP) for road network operation and maintenance, and the percentage of trips made by public transport, some Eastern European cities (Moscow, Prague, Warsaw, Krakow, etc.) appear to be the most sustainable. However, these patterns are likely not the result of particularly efficient public transport systems. Instead, they stem from the almost unavoidable reliance on public transport, given that purchasing, operating, and maintaining private cars entails costs that are disproportionately high relative to the economic conditions of these countries.

Considering variables such as the percentage of trips made by private vehicle, annual motorized passenger kilometers per capita, and the ratio of annual motorized private transport passenger kilometers to those of public transport, Western European cities emerge instead, such as Munich, Marseille, Copenhagen, and other Northern European cities. Continuing the analysis, it was also observed that for other sustainability-related variables—such as road accident rates, motorization rate, energy consumption, and pollutant emissions (all negative indicators from a sustainability perspective)—higher values are consistently found in more developed metropolitan areas, especially in Italian cities such as Rome, Turin, and Bologna. In general, major Italian cities show a significant gap in sustainability compared to the average of other large European cities.

Differences in sustainability values become even more pronounced when comparing cities from different parts of the world. These differences are mainly due to fundamentally different urban development patterns and land-use characteristics.

Newman and Kenworthy (1999) and Kenworthy and Laube (1996) identified and compiled a large set of indicators for major cities across several continents, analyzing the relationship between factors associated with urban transport, land-use patterns, and car use. The results reveal a significant difference in car use between the US and Australian cities compared with European and Asian ones: even though European cities (at the time of the study) had a per capita income between 20 and 50% higher than that of Australian and US cities, car use was far less prevalent. The difference is even more pronounced when comparing the United States and Australian cities with Asian ones. The reason for this difference lies, as noted, in the profound disparities in urban form. By analyzing the relationship between car use, public transport use, and urban configuration, expressed through density, strong correlations emerge. As density increases, car use decreases sharply, while public transport use increases systematically.

In summary, various factors influence the sustainability level of mobility systems. Urban configuration is certainly fundamental, yet difficult to change both operationally and in terms of time. Therefore, strategies to improve sustainability in the short to medium term must necessarily involve the application of appropriate policies concerning both supply and demand.

3.2 TRANSPORT POLICIES FOR SUSTAINABLE MOBILITY

There are many policies available to improve the sustainability of a transport system. These include land-use policies, policies aimed at changing users' attitudes and behaviors, policies for the design and management of transport infrastructure, policies related to the dissemination of transport system information, pricing policies, logistics, particularly urban logistics, and policies concerning system safety, to name the most important categories. Thus, several solutions are available to decision-makers. The challenge lies in identifying which policies are most suitable for each specific context, bearing in mind that transport systems, particularly in medium and large cities, are extremely complex.

These systems serve a multitude of users, each with their own mobility needs, and are composed of several subsystems: automobile traffic, motorcycle traffic, public transport, cycling, and walking. These subsystems are also interconnected: if a problem occurs, for example, on a metro line, the effects are typically reflected on the road system, where congestion can arise, and/or on surface public transport.

This aspect requires the implementation of integrated packages of measures capable of managing not only the potential effects of each intervention on the targeted subsystem, but also on the others. In the following sections of the chapter, attention will focus on two categories of policies that are particularly important in transport: land-use policies and pricing policies, examining their characteristics and their effects on sustainability. At the end of the chapter, several tools to support decision-makers in planning transport system sustainability are presented.

3.2.1 Land-use policies

We have seen the relationship between urban form and car use. Figure 3.2 shows how car ownership (motorization rate) has changed in Italian metropolitan areas between 2000 and 2017. The variation in the motorization rate in the central municipality of the metropolitan area is compared with that of its surrounding belt, that is, what occurred in the hinterland of the main municipality.

If, for example, we look at the case of Rome, we observe a decrease in the motorization rate in the central municipality and an increase in the metropolitan belt. Setting the 2000 value of the rate to 100, the 2017 value is 95.1, a reduction of about 5% in the number of cars per inhabitant. In the metropolitan belt, the opposite occurs, with an increase of 7.5%. With a few exceptions, this phenomenon can be observed in most Italian metropolitan cities.

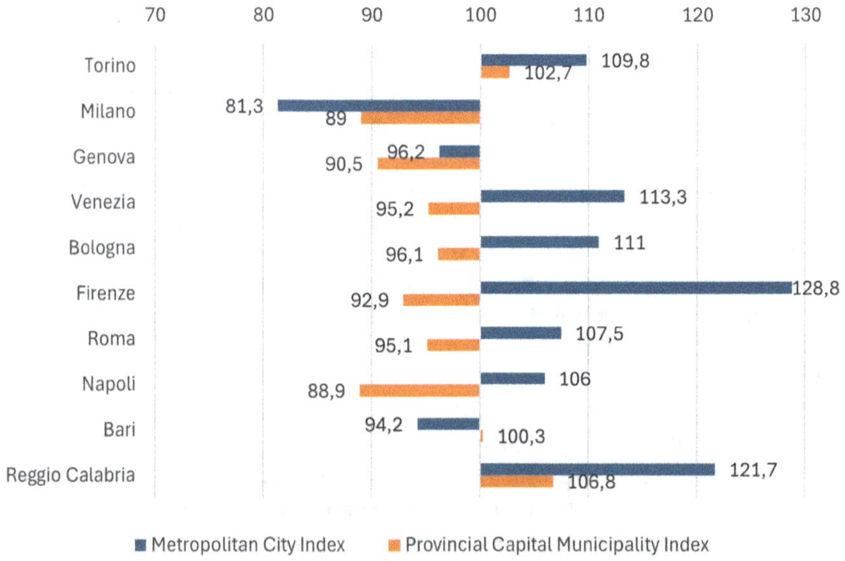

Figure 3.2 Change in the motorization rate in metropolitan areas in the period 2000–2017.

Source: Authors' own elaboration based on Isfort (2018).

3.2.1.1 Density and mixed land use

How can urban development be guided so that it promotes the sustainability of the transport system? It is necessary to act on those key land-use variables that influence travel behavior and therefore transport demand.

Two important variables are density and mixed land use.

Density refers to the number of activities, such as dwellings, households, people, and jobs per square kilometer (or any other surface-based indicator). Where such densities are higher, people can reach the places they need for their daily activities by traveling shorter distances, thus saving time and money. Moreover, thanks to shorter distances, more destinations can be reached on foot or by bicycle, encouraging a shift from cars to active modes of transport.

Mixed land use refers to the degree of integration of different categories of land use, such as housing, workplaces (businesses), shops, schools, and medical services. Similar to the effects of higher density, a greater level of mixed land use can improve accessibility and therefore reduce the need for car travel.

3.2.1.2 Land-use planning to support public transport

Urban development should also be oriented toward supporting the use of public transport. Encouraging the use of public transport through land-use planning involves locating new developments and managing existing ones in such a way as to:

- Improve the conditions for an efficient operation of public transport;
- Improve the accessibility of urban areas and allow people to travel more using alternative modes;
- Increase demand for public transport, particularly by encouraging a modal shift away from private cars.

In this sense, a key factor is the distance from public transport stops. The distance from railway stations can have a significant impact on modal choice, as differences in travel times by car and by train are highly influential. For public transport, total travel time includes both "in-vehicle time" (e.g., on the train) and access and egress time (for example, traveling from home to the station by bus, bicycle, or on foot, and from the station to the office by walking or taking local public transport). If a greater number of homes and workplaces are located near railway stations, access and egress times are shorter for more users, thereby favoring the use of trains over cars.

The general principle is therefore to ensure that the origins and destinations of trips are situated along nodal or linear networks compatible with demand to ensure that public transport services, both bus and rail, are feasible and efficient. Moreover, major activities, employment centers, and high-density residential areas should be encouraged near stations, major stops, and interchanges along public transport corridors.

An example is the case study of Stockholm, where, thanks to careful planning over the past 60 years, an urban configuration ideal for public transport use has been achieved. During the expansion phase of Stockholm's development in the 1950s and 1960s, the city became an example of integrated planning of new neighborhoods and new infrastructure. Districts such as Vällingby and Farsta were developed with densities specifically calibrated to sustain a new metro line, which was completed precisely as residents and businesses moved in.

Crucially, the development of workplaces along rail corridors was encouraged, resulting not only in easy access to these job centers via public transport but also in balanced travel flows. During peak hours, 55% of commuters travel in one direction and 45% in the opposite direction, which greatly reduces overloads on the network. This has supported high levels of public transport use, with modal shares that in some cases are nearly double those of other major European cities such as Berlin and London.

3.2.1.3 Parking standards

Among land-use measures, parking standards are among those that most directly influence levels of car use. These standards consist of regulations establishing the minimum, and sometimes maximum, number of parking spaces required or permitted for specific types of new developments within a given area.

Similar to on-street parking regulation and pricing, parking standards are applied both to meet demand and to manage supply in order to influence travel behavior.

This approach has been increasingly criticized for having led, in recent years, to an oversupply of parking and a rise in car trips. It relies on estimates of peak parking demand, which tend to push the provision of private parking upward and effectively reduce usable land and development density—thereby exacerbating urban sprawl and car dependency. In a vicious cycle, the resulting oversupply of parking further encourages driving and ultimately leads to levels of parking demand even higher than originally forecast.

A different approach is the introduction of maximum parking standards, which are designed primarily to reduce car use. However, this approach is effective only where viable alternatives to the private car exist; otherwise, it risks creating significant accessibility problems.

3.2.1.4 Developer contributions

A final example of land-use policy is that of developer contributions, which consist in requiring developers to finance the construction of transport infrastructure in the area where they build.

Such contributions are justified both because the presence of a new development places additional strain on existing infrastructure by attracting or generating new trips and because the businesses that will locate in the area will benefit from the transport infrastructure already in place.

Some effects of this measure, particularly those related to the area's economic development, remain unclear. Although there is broad agreement among policymakers and businesses that transport infrastructure can play a significant role in supporting economic growth, there is still no definitive evidence of the economic benefits that might be achieved.

3.2.2 Pricing policies

Transport policies can be divided into two broad categories: push policies and pull policies (often referred to in Italian as "bastone e carota"). Push policies aim to "push" users away from private car use through disincentive measures. Pull policies, instead, aim to "pull" users toward more sustainable modes such as public transport, cycling, and walking by making these options more attractive.

Push measures tend to achieve the most significant results, but they are generally unpopular. Pull measures are more acceptable to users, but they usually shift only limited shares of travelers. The most effective strategies combine both approaches.

Among push measures, pricing policies targeting the use, and in some cases the ownership, of private vehicles are particularly effective. The most important ones are described below.

3.2.2.1 Vehicle ownership tax

Vehicle ownership taxation (an indirect tax) has two main purposes. First, to generate general government revenue; second, to regulate the number of vehicles owned and, indirectly, the age of the vehicle fleet in order to meet environmental objectives.

There are also differences across vehicle types: for example, electric vehicles in Italy do not pay the ownership tax for the first 5 years, and similarly, hybrid, methane, and LPG (Liquefied Petroleum Gas) vehicles may benefit from exemptions or reductions in many regions.

3.2.2.2 Fuel tax

Fuel represents the highest operating cost for motor vehicles. Increasing vehicle operating costs tends to reduce private car use. For this reason, fuel tax increases are sometimes proposed to reduce car use and improve the efficiency of the transport system.

In practice, this functions as a distance-based tax: the more kilometers one drives, the more one pays. This aligns with the pay-per-drive principle, justified by the fact that driving more kilometers generates more pollutant emissions, noise, and road accidents, as well as greater wear on road infrastructure. Ideally, such a tax enables the state to compensate for the externalities and damages caused by private vehicle use.

3.2.2.3 Parking pricing

Parking pricing means that motorists pay directly for the use of parking facilities. It can be implemented as a strategy to reduce vehicle traffic, to manage parking demand (reducing parking-related problems), to recover the costs of providing parking infrastructure, to generate revenue for other purposes, or to pursue a combination of these objectives.

3.2.2.4 Road pricing

Road pricing, also known as *congestion charging* or *road user charging*, requires motorists to pay to drive on a specific road or within a specific area during certain times. The main goal is to limit private car use by reducing congestion and, consequently, environmental impacts.

This type of policy must be accompanied by parallel interventions on other subsystems (public transport, cycling, and walking) to prevent the priced area from becoming less accessible, which would negatively affect economic activity.

A well-known example in Italy is Milan's Area C, while the most internationally renowned case is London's Congestion Charging Scheme, which has significantly improved the sustainability of the central area of the UK capital without producing negative effects on economic activity or on surrounding districts.

3.2.3 International databases

The previous sections described several categories of policies aimed at improving sustainability, but the range of available solutions is very broad. There are various international databases that describe these policies in detail, highlighting their applicability to different contexts, their effectiveness, efficiency, and the barriers to implementation.

KonSULT (http://www.konsult.leeds.ac.uk/) is a database developed with support from the European Commission. It is designed to help policymakers, practitioners, and stakeholders understand the challenges involved in achieving sustainability in urban transport and to identify suitable policies and policy packages. It also provides detailed information relevant to researchers and students.

The Transport Demand Management (TDM) Encyclopedia (https://www.vtpi.org/tdm/index.php), produced by the Victoria Transport Policy Institute, aims to enhance the understanding and implementation of TDM and serves as a comprehensive source of information on innovative solutions for managing transport problems. It provides detailed information on dozens of demand management strategies, as well as general information on TDM planning and evaluation techniques.

ELTIS (www.eltis.org) is an EU platform that provides information, good practices, tools, and communication channels to support the transition toward a sustainable urban mobility model. In particular, the section dedicated to mobility plans serves as an information hub on how to develop and implement sustainable urban mobility plans (SUMPs).

3.3 INNOVATION AND FUTURE TRENDS

Road transport is a major source of pollutant emissions. As our demand for mobility and transport continues to grow, we are placing increasing pressure on nature's capacity to sustain this expansion. Innovation can provide solutions to meet the mobility needs of future generations while simultaneously improving sustainability.

Innovation in vehicle design and manufacturing can promote the widespread adoption of highly energy-efficient, low-emission vehicles. Innovation in mobility services, supported by the digital economy, will simplify travel and encourage the use of assets such as private cars over ownership.

3.3.1 Electric mobility

3.3.1.1 What is electric mobility?

One solution that clearly addresses the requirements of sustainable mobility is electric mobility, which can be defined as a road transport system based on electrically powered vehicles whose energy is supplied by an external source of electricity (Sandén & Wallgren, 2017). This definition requires the adoption of systems thinking in order to fully understand the advantages and disadvantages of the various electric vehicle technologies. A transport system that uses electricity from the grid can, in fact, draw energy from many different sources without major modifications to electric vehicles or energy supply systems. However, if vehicles use electricity generated from coal, the climate impacts of electric propulsion may be worse than those of gasoline- or diesel-powered vehicles (Figure 3.3).

3.3.1.2 What are the technologies for electric mobility?

The main propulsion configurations for electric mobility are battery electric vehicles, continuously powered electric vehicles, fuel cell vehicles, and hybrid vehicles.

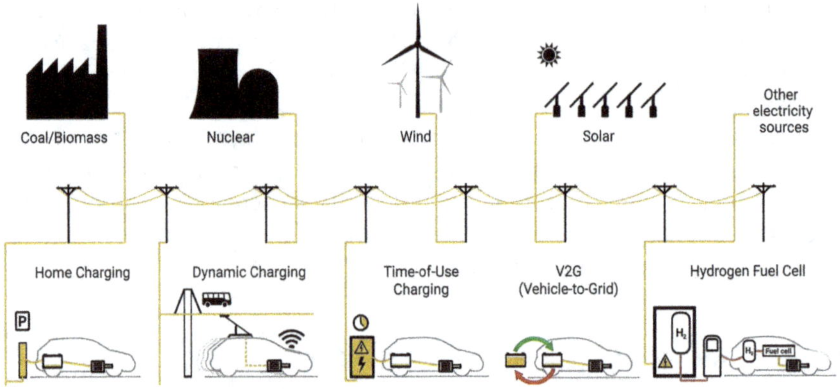

Figure 3.3 Sources and technologies of electric mobility.

Source: Sandén & Wallgren, 2017.

Battery electric vehicles (BEVs). In these vehicles, electrical energy is stored in a rechargeable battery. The direct current from the battery is converted into alternating current by inverters to drive the electric motor. The electric motor then converts electrical energy into mechanical energy. At present, the three main barriers perceived by consumers are purchase costs, battery charging times (a standard charge can take up to 8 hours, which is a major limitation when traveling), and so-called range anxiety, that is, anxiety related to uncertainty about driving range. Several options are available to extend the driving range of electric vehicles. The most common is fast charging; another solution is rapid battery swapping.

Continuously powered electric vehicles. Vehicles with a continuous power supply draw energy directly from the electricity grid while in motion, thereby reducing the need to store energy onboard and effectively overcoming the range limitation. In this case, the main obstacle is the construction of infrastructure integrating conductive power lines or inductive rails, which requires substantial investment.

Fuel cell vehicles (FCVs). A fuel cell is an electrochemical device that converts the chemical energy of a fuel (often hydrogen) and an oxidizing agent (often oxygen) into electricity through a pair of redox reactions that are more efficient than batteries. In principle, a fuel cell operates similarly to a battery: it stores energy from external sources and releases it when needed. Hydrogen outperforms batteries in terms of energy capacity and refueling time. However, hydrogen technology is currently much more expensive due to the higher cost of hydrogen vehicles compared to conventional vehicles, as well as the need to develop production and distribution infrastructure.

Plug-in hybrid electric vehicles (PHEVs). Due to the limitations of the technologies described above, hybrid solutions are available on the market, as combining two powertrains allows the strengths of each to be exploited while compensating for their weaknesses. A plug-in hybrid electric vehicle is a vehicle whose battery can be recharged by connecting it to an external source of electricity, in addition to being charged by its onboard engine and generator.

3.3.1.3 What are the sustainability impacts of electric mobility?

To assess the impacts of electric mobility, Well-to-Wheel (WTW) studies are commonly used. This type of life cycle assessment (LCA) focuses on the life cycle of the energy carrier used to power the vehicle, whether liquid fuel or electricity. The impact associated with vehicle operation is considered in the phase in which the energy carrier is used to propel the vehicle, known as tank-to-wheels (TTW). The preceding phase, known as well-to-tank (WTT), focuses on energy supply to the vehicle. The environmental burden of the WTT phase varies depending on how the energy carrier is produced. For example, historical data from 2008 based on an average EU electricity mix of 467 g CO_2-eq/kWh highlighted a significant difference between electricity generated by hydropower plants and that produced by coal-fired power plants (Sandén & Wallgren, 2017).

A WTW analysis is performed by linking the WTT and TTW phases. A complete LCA also includes the vehicle life cycle itself, namely vehicle production and end-of-life disposal. Recent assessments show a significant evolution in the climate benefits of electromobility compared to earlier decades. While older analyses already suggested that greenhouse gas (GHG) emissions per kilometer for BEVs in Europe were lower than those of internal combustion engine vehicles (ICEVs), recent findings indicate that this gap has widened significantly due to the rapid decarbonization of the energy grid (Negri & Bieker, 2025).

According to the latest data, medium-segment BEVs registered in Europe in 2025 produce 73% fewer life cycle GHG emissions than comparable gasoline-powered cars (Negri & Bieker, 2025). Specifically, while a gasoline ICEV emits approximately 235 g CO_2 e/km, a BEV charging with the projected average EU electricity mix (2025–2044) emits only 63 g CO_2 e/km. If powered entirely by renewable electricity, BEV emissions drop further to 52 g CO_2 e/km, representing a 78% reduction compared to gasoline vehicles (Negri & Bieker, 2025).

In terms of vehicle categories, current studies differentiate between hybrids (HEVs), plug-in hybrids (PHEVs), and BEVs. PHEVs typically offer a 30% reduction in lifetime emissions compared to gasoline cars, while HEVs achieve a more modest 20% reduction. The E-REV category (extended-range electric vehicles), powered by an electric motor with an internal combustion engine used solely as a generator, occupies an intermediate position in this hierarchy.

By integrating the vehicle life cycle into the analysis, it is observed that emissions associated with production processes remain higher for electric vehicles—approximately 40% higher than for ICEVs—due to the energy-intensive mining and processing of materials for batteries. However, this initial "emissions debt" is now offset much earlier than previously estimated. In the European context, the higher production emissions of a BEV are typically offset after only 17,000 km of driving, usually within the first one to two years of use (Negri & Bieker, 2025). Consequently, the overall CO_2 balance of electric vehicles remains overwhelmingly positive across all major markets, including Europe, the United States, and China, provided the transition to renewable energy sources continues as projected.

3.3.2 Shared mobility

3.3.2.1 What is shared mobility?

Shared mobility refers to various modes and services that increase the efficiency of the transport system through vehicle sharing. Examples include ride-hailing services (such as Uber and Lyft), car sharing, and bike sharing.

New internet and telecommunication technologies facilitate the use of these modes, for example, through dedicated platforms that connect supply and demand for these services.

When classifying shared mobility services, four main reference models can be identified (Figure 3.4): peer-to-peer car rental, car sharing, ride sharing, and carpooling.

Below is an overview of the four main vehicle-sharing models.

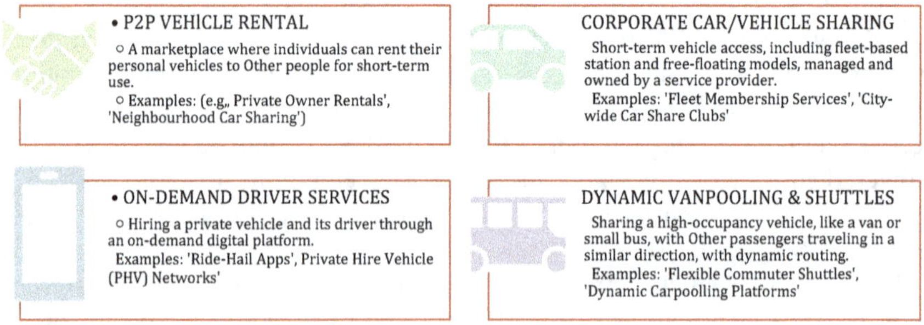

Figure 3.4 Mobility service models.

Source: authors' own elaboration based on Santos, 2018.

3.3.2.2 Model 1: Peer-to-peer vehicle rental

The idea behind private car sharing, or peer-to-peer (P2P) vehicle rental, is that private car owners make their vehicles available for rental to neighbors and the general public when they do not need them. This allows car owners to earn income from their vehicles and reduce the costs of car ownership. For people who do not own a car, the scheme offers convenient access to a vehicle when needed, often within the same neighborhood or even next door.

The relationship between owner and user is mediated through an online platform (app or website) managed by an operator, while vehicle rental procedures and insurance coverage are typically handled by the operating company. This service is beginning to spread in Italy, where one platform is currently available (auting.it). Among the barriers, regulatory constraints may arise; for example, in Italy, vehicles used for rental must be registered for third-party use.

3.3.2.3 Model 2: Corporate/operator-managed car sharing (station-based and free-floating)

Car sharing is a form of short-term vehicle rental in which users can rent vehicles on a pay-per-use basis or through a monthly or annual membership fee. Vehicles are distributed across designated areas, forming stations, from which registered members can pick them up either immediately or after making a reservation through a platform. This traditional service is also referred to as station-based car sharing.

By contrast, in free-floating services, vehicles can be picked up and dropped off freely within a predefined area. No stations are required; vehicles are equipped with GPS and can be located by users through an app.

More generally, the term "vehicle sharing" includes services in which the shared vehicle may be a bicycle (bike sharing), a moped (scooter sharing), or an electric scooter. In Italy, these services have increasingly become an integral part of multimodal transport systems.

3.3.2.4 Model 3: Ridesourcing/transportation network companies (TNCs), e-hailing

Companies such as Uber and Lyft are providers of ridesourcing services (although they are often labeled as ridesharing providers). In these services, trips are requested by one or more users in exchange for payment for a commercial service that includes both driving activity and vehicle provision.

These companies are often referred to as transportation network companies (TNCs). TNCs offer on-demand private drivers providing door-to-door services, typically activated via smartphone applications. Well-known examples include Uber and Lyft. This includes pooled options (e.g., UberPOOL) such as ridesplitting, where multiple users share the same driver-driven vehicle at a reduced rate.

3.3.2.5 Model 4: carpooling and vanpooling (dynamic ridesharing) + commuter shuttles

This is a mobility service based on the shared use of private vehicles by two or more people traveling along the same route, or part of it. Carpooling typically involves cars carrying up to 7–8 passengers, while vanpooling uses minibuses that can transport between 7 and 15 passengers. Travel costs are shared between the driver and passengers.

These services can be further classified according to geographical scope and user type, including:

- Urban carpooling (e.g., CLACSOON);
- Interurban or medium- to long-distance carpooling (e.g., BlaBlaCar);
- Corporate carpooling dedicated to home-to-work commuting.

3.3.2.6 What are the sustainability impacts?

Shared mobility has mixed impacts on travel behavior, depending on the type of service and how it is used. Research results in this area are still limited.

Some shared mobility options, such as e-hailing (Model 3) and car sharing (Model 2), involve car travel and may therefore contribute to an increase in the number of car trips. At the same time, however, they can reduce the need for households to own a car, thereby contributing to a reduction in total car travel. In the Netherlands, a survey of car-sharing and P2P car rental users (Models 1 and 2) found a 30% lower car ownership rate among car-sharing users and a 15–20% reduction in kilometers traveled compared to the period before adopting car sharing. Shared vehicles mainly replaced a second or third household car (Santos, 2018). The study also showed that, as a result of reduced car ownership and use, car-sharing users emitted less CO_2 per person per year.

Bike sharing can reduce car use by providing an alternative for specific trips (e.g., last-mile travel) and by adding transport service that reduces the need for privately owned vehicles.

In the United States, the Uber/Lyft model does not appear to have had the same impact on the number of cars per million inhabitants. The share of total paid car trips made using Uber has reached approximately 50% in several US cities, reflecting a shift primarily away from taxis rather than from private cars.

The introduction of Uber/Lyft-type services (Model 3) in US cities has increased vehicle ownership by an average of 0.7% (Ward et al., 2021). This increase is larger in urban areas with higher initial motorization rates (car-dependent cities) and in areas with low population growth. On average, there is no statistically significant effect on public transport use, although these services tend to substitute public transport more in cities where potential users have higher willingness and ability to pay for more expensive and convenient travel modes and where users are more likely to be without children and therefore face fewer family-related constraints, such as the need for child seats (Ward et al., 2021). The growing market share of these new mobility services may therefore primarily reflect a shift away from taxis rather than from private cars. In several US cities, the share of paid car trips carried out by Uber has reached around 50%, mirroring a decline in the share of business trips made by taxis (Santos, 2018).

A comparison of CO_2 emissions generated by carpooling trips on the BlaBlaCar platform (Model 4) with those that would have occurred without the platform shows a 26% reduction in CO_2 emissions (Le BIPE, 2018). A simulation conducted for the city of Helsinki showed that, assuming only 20% of car and taxi trips are replaced by shared mobility (Model 4), vehicle kilometers traveled would decrease by 7%, CO_2 emissions by 2%, and congestion by 13% (Santos, 2018).

3.3.3 Cooperative, connected, and automated mobility

Cooperative, connected, and automated mobility (CCAM) represents the future of sustainable transport planning.

The challenge for automotive manufacturers is to develop connected vehicles supported by advanced technologies and artificial intelligence, such as autonomous vehicles, capable of significantly improving the quality of life while also preserving environmental resources.

Connected vehicles use vehicle-to-everything (V2X) communication technologies to interact with drivers, other vehicles, and infrastructure. These include:

- V2P (vehicle-to-pedestrian): enables vehicles to detect the presence of vulnerable road users;
- V2V (vehicle-to-vehicle): allows vehicles to exchange wireless information on speed, position, and direction;
- V2I (vehicle-to-infrastructure): represents the next generation of Intelligent Transport Systems (ITS);
- V2D (vehicle-to-device): involves the exchange of information between a vehicle and any electronic device connected to it;
- V2N (vehicle-to-network): enables broadcast and unicast communication between vehicles and the V2X system (e.g., warnings about road accidents, congestion, or queues along the planned route).

These technologies can be used not only to improve vehicle safety but also to enhance vehicle efficiency and reduce travel times.

An autonomous vehicle, by contrast, uses a combination of sensors, cameras, radar, human–machine interfaces (HMI), and artificial intelligence (AI) to travel between destinations without the need for human intervention. The Society of Automotive Engineers (SAE) has adopted a classification scale from Level 0 to Level 5 to assess the degree of vehicle automation (Figure 3.5), where Level 0 corresponds to a traditional vehicle with no driver assistance, and Level 5 represents full automation, requiring only destination input and system activation by the user, with no further human intervention. Intermediate levels include advanced driver assistance systems (ADAS) and partial vehicle autonomy.

The potential benefits that automated and connected vehicles can bring to the sustainability of the transport system are significant from both environmental and social perspectives. These advantages include reduced pollution, improved efficiency resulting from smoother driving behavior—such as fewer abrupt braking and acceleration events—as well as increased productivity and comfort. In addition, automated and connected vehicles may enhance mobility inclusion for specific groups, such as people with disabilities or older adults.

Other potential benefits, however, remain highly uncertain. For example, CCAM technologies are expected to reduce road accidents by eliminating human error; nevertheless, during the transition period in which a mix of conventional and automated vehicles will share the same infrastructure, their introduction may actually increase risk. Moreover, the growth in transport demand induced by the deployment of CCAM could lead to higher levels of traffic congestion, with serious consequences for the environment and for the mobility of both people and goods.

Within the ADAS&ME project, funded by the European Commission under the Horizon 2020 program, an impact assessment was conducted on specific ADAS designed to monitor and adapt to the driver's psychophysical state—such as fatigue/drowsiness, stress, inattention/distraction, and impaired emotional states—as well as to environmental conditions. This was achieved using human–machine interface (HMI) systems and automated driving systems across different vehicle types, including conventional and electric cars, trucks, buses, and motorcycles. The study focused on the analysis of seven use cases, illustrated in Figure 3.6, which combine the vehicle types with different driver states.

By way of example, the results of the impact assessment for Use Case G are presented. This use case concerns the monitoring of conditions such as stress and drowsiness during the transition to autonomous driving mode of a bus while passengers are boarding and alighting at a stop, followed by a return to manual driving once the system has detected that the driver is fit to resume control.

Level	Name	Narrative definition	Execution of steering and acceleration/ deceleration	Monitoring of driving environment	Fallback performance of *dynamic driving task*	System capability (*driving modes*)
***Human driver* monitors the driving environment**						
0	No Automation	the full-time performance by the *human driver* of all aspects of the *dynamic driving task*, even when enhanced by warning or intervention systems	Human driver	Human driver	Human driver	n/a
1	Driver Assistance	the *driving mode*-specific execution by a driver assistance system of either steering or acceleration/deceleration using information about the driving environment and with the expectation that the *human driver* perform all remaining aspects of the *dynamic driving task*	Human driver and system	Human driver	Human driver	Some driving modes
2	Partial Automation	the *driving mode*-specific execution by one or more driver assistance systems of both steering and acceleration/deceleration using information about the driving environment and with the expectation that the *human driver* perform all remaining aspects of the *dynamic driving task*	**System**	Human driver	Human driver	Some driving modes
***Automated driving system* ("system") monitors the driving environment**						
3	Conditional Automation	the *driving mode*-specific performance by an *automated driving system* of all aspects of the *dynamic driving task* with the expectation that the *human driver* will respond appropriately to a *request to intervene*	System	**System**	Human driver	Some driving modes
4	High Automation	the *driving mode*-specific performance by an *automated driving system* of all aspects of the *dynamic driving task*, even if a *human driver* does not respond appropriately to a *request to intervene*	System	System	**System**	Some driving modes
5	Full Automation	the full-time performance by an *automated driving system* of all aspects of the *dynamic driving task* under all roadway and environmental conditions that can be managed by a *human driver*	System	System	System	**All driving modes**

Figure 3.5 Automated driving levels.

Source: SAE International/Wikimedia Commons, 2021. Licensed under Creative Commons Attribution-ShareAlike 4.0 International [CC BY-SA 4.0].

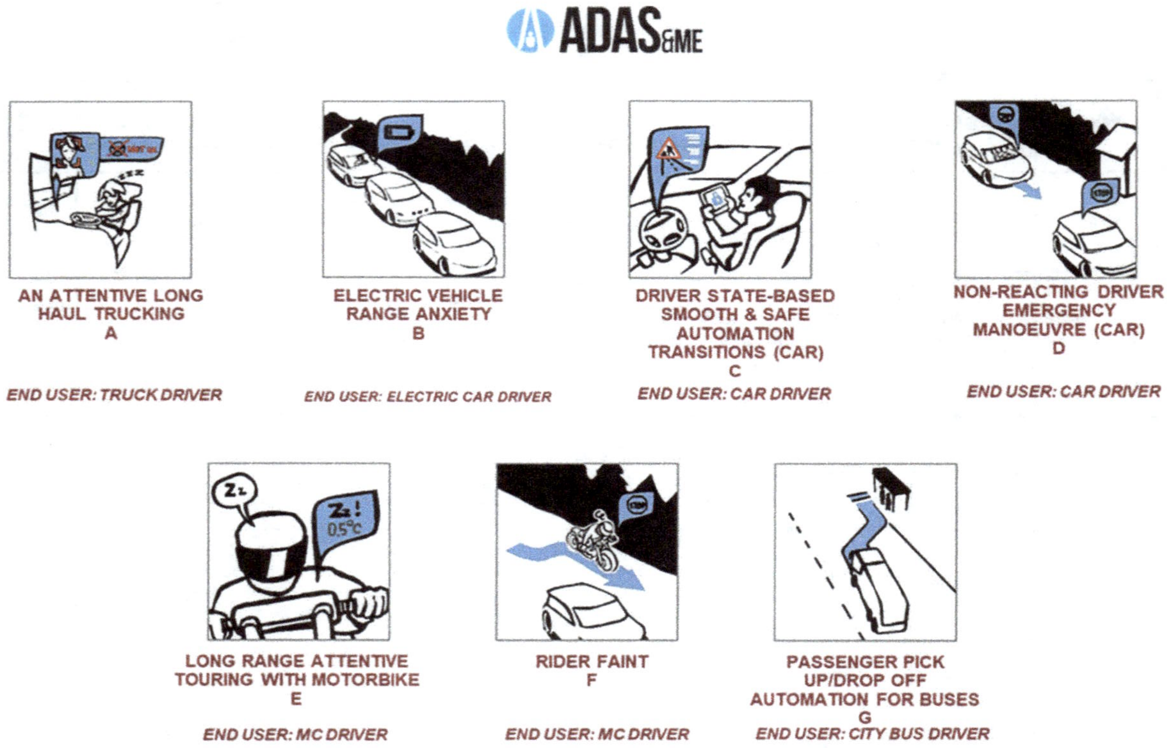

Figure 3.6 ADAS&ME use cases.

Source: Meta et al. (2020), licensed under CC BY 4.0.

Regarding safety-related aspects, the results show that all ADAS&ME systems are able to mitigate the negative effects of adverse driving states—which can increase accident risk—and to enhance driving comfort. Moreover, in the specific case of Use Case G, improvements in longitudinal and lateral acceleration during the autonomous driving phase were observed to the benefit of passenger comfort. Concerning mobility-related aspects, a reduction in travel times was identified, with a consequent decrease in energy consumption, due to the optimization of stopping phases.

3.4 TRANSPORT PLANNING TOOLS

As discussed, there is a wide range of policies aimed at improving the sustainability of transport systems. At the same time, such policies should never be implemented in isolation, but rather as part of structured policy packages designed to address the various subsystems, thereby avoiding situations in which improvements in one area lead to deterioration in others. Transport planning tools are, in fact, integrated plans that start with the analysis of existing problems, followed by the identification of objectives, the selection and evaluation of alternative solutions, and finally their implementation and monitoring.

There are several types of plans, both general—such as traffic plans, SUMPs, regional transport plans, and home-to-work or home-to-school travel plans—and sector-specific plans, including public transport plans, urban logistics plans, and road safety plans. Two particularly relevant examples are discussed below: the SUMP and the Home-to-Work/Home-to-School Travel Plan, the latter illustrated through the specific case of the Sustainable Mobility Plan of Sapienza University of Rome.

3.4.1 Sustainable urban mobility plan (SUMP)

The Sustainable Urban Mobility Plan is a strategic planning instrument designed to contribute to the achievement of climate and energy targets set at the European level. Within the broader planning framework, it acts as a coordinating tool that interfaces with other plans and programs aimed at promoting increasingly sustainable forms of mobility.

In recent years, the approach to transport planning has changed significantly, both in academia and in practical planning. Compared to traditional approaches (Figure 3.7), the new paradigm proposed by SUMP places people at the center of the process. The main objectives are accessibility, economic feasibility, social equity, and health. Short- to medium-term planning is embedded within a long-term strategic vision and must account for the development of all modes of transport while simultaneously encouraging the use of sustainable ones.

Traditional Transport Planning		Sustainable Urban Mobility Planning
Focus on traffic	→	Focus on **people**
Primary objectives: Traffic flow capacity and speed	→	Primary objectives: **Accessibility** and **quality of life**, including social equity, health and environmental quality, and economic viability
Mode-focussed	→	**Integrated development of all transport modes** and shift towards sustainable mobility
Infrastructure as the main topic	→	**Combination** of infrastructure, market, regulation, information and promotion
Sectoral planning document	→	Planning document **consistent with related policy areas**
Short and medium-term delivery plan	→	Short and medium-term delivery plan embedded in a **long-term vision and strategy**
Covering an administrative area	→	Covering a **functional urban area** based on travel-to-work flows
Domain of traffic engineers	→	**Interdisciplinary** planning teams
Planning by experts	→	Planning with the **involvement of stakeholders and citizens** using a transparent and participatory approach
Limited impact assessment	→	Systematic **evaluation** of impacts to facilitate **learning** and improvement

Figure 3.7 Differences between traditional transport planning and sustainable urban mobility planning.

Source: Rupprecht Consult (ed.), 2019.

The preparation of a SUMP involves interdisciplinary working groups, includes citizen engagement through transparent and participatory processes, and is coordinated with planning documents from other related sectors. Finally, the new approach places strong emphasis on continuous, structured monitoring over time, as well as on impact assessment of actions and interventions, in order to identify shortcomings, correct them, and achieve continuous improvement.

At the national level, Ministerial Decree No. 397 of August 4, 2017 sets out the guidelines for the preparation of SUMPs. The Decree identifies the SUMP as the primary mobility planning instrument at the municipal and metropolitan city level, followed by the urban traffic plan (PUT). It defines:

- The SUMP, a medium- to long-term strategic plan that addresses mobility issues whose resolution requires investment and therefore financial resources and technical implementation time, as well as the development of complex, intersectoral urban, and metropolitan policies. The objectives are pursued without assuming unchanged infrastructure resources.
- The PUT (urban traffic plan), a short-term plan that assumes unchanged infrastructure resources and focuses on the optimal organization of existing assets; it is therefore essentially a management plan.

In encouraging a uniform procedure for the drafting and approval of SUMPs, the Decree provides, among its various indications:

- A possible structure for a timeline including all procedural steps to be followed, from the definition of the working group to the formal adoption of the SUMP.
- An example of a SUMP table of contents, further detailing the working phases;
- A list of objectives, strategies, and actions that the SUMP is required to pursue and achieve;
- A list of indicators for monitoring activities that must be activated following approval of the SUMP.

Ministerial Decree No. 396 of 28 August 2019 introduces amendments to Decree No. 397 of 4 August 2017, while leaving the overall structure of the guidelines unchanged.

National legislation on SUMPs follows the recommendations provided by the European Commission (EC) in the Guidelines for Developing and Implementing a Sustainable Urban Mobility Plan, published in January 2014 and updated in 2019 (Rupprecht Consult, 2019), which are available in nine languages.

The guidelines identify four phases for the development and implementation of a SUMP. Each phase is characterized by three steps and a milestone (Figure 3.8).

Phase 1: Preparation and Analysis consists of the following steps: establishing working structures, defining the planning framework, and analyzing mobility conditions. At the end of this phase, the results of the analysis must be shared in a baseline report, including all identified problems and opportunities.

Phase 2: Strategy Development includes the following steps: building and jointly assessing scenarios; developing a shared vision and strategy with stakeholders; and defining objectives and indicators. The outcomes of this phase must be consolidated into a summary document, and before moving to the next phase, consideration should be given to gathering feedback from citizens on the identified strategic priorities.

Phase 3: Measure Planning consists of the following steps: selecting packages of measures together with stakeholders; agreeing on actions and responsibilities; and preparing for adoption and funding. The most important milestone of this phase is the formal adoption of the Sustainable Urban Mobility Plan by political decision-makers. The adoption process may take several months and depends on both the national regulatory framework and that of the competent authority. Once adopted, the plan should be publicly presented during a final celebratory event involving stakeholders, the local community, and the media.

Figure 3.8 Steps of sustainable urban mobility planning (2nd edition of the guidelines).

Source: Rupprecht Consult (ed.), 2019.

As with the previous phases, **Phase 4: Implementation and Monitoring** is also structured into three steps: coordinating and managing implementation, monitoring progress while informing and engaging citizens and stakeholders, and analyzing successes and failures. This final phase enables reflection on the work carried out and supports continuous improvement in future planning processes.

3.4.2 Case study: The sustainable university mobility plan of Sapienza University of Rome

The sustainable university mobility plan of Sapienza University of Rome (hereinafter referred to as the SUMP) is framed within a broader international context of universities committed to pursuing environmentally sustainable mobility policies (CTL, 2020). It originated from the University's intention, in collaboration with the Centre for Transport and Logistics Research (CTL), to understand and improve—toward greater sustainability—the travel patterns of students and staff at one of the oldest and largest universities in the world, whose campuses are located within the complex and challenging urban context of the city of Rome.

The development of the SUMP followed a lengthy process that included two key phases:

- Survey and analysis of student and staff travel behavior;
- Consultation with a panel of experts and the formulation of proposals for sustainable university mobility.

For the survey, two questionnaires were designed: one addressed to students and the other to staff. The questionnaires were administered online using an advanced, purpose-built digital platform. Based on respondents' answers regarding their travel modes, the platform displayed only the questions relevant to each individual case. The questionnaire, sent to 107,340 students and 23,101 staff members, included questions on:

- Personal information, role, and place of study or work;
- Travel characteristics: home–university, university–home, and trips between different campuses;
- Information and motivations influencing the choice of the main travel mode—car, motorcycle or scooter, public transport, bicycle—including costs, ownership, and willingness to adopt more sustainable mobility options;
- Perceptions and attitudes related to travel;
- General information, also related to road safety.

Approximately 14% of students (14,719 respondents out of 107,340) and 41% of staff (9403 respondents out of 23,101) completed the questionnaire. A summary of the characteristics of the respondent sample shows that:

- Female respondents were more engaged with the initiative, participating at a higher rate than males and providing a larger share of responses;
- Half of the students and 70% of staff are highly engaged with the university environment, traveling to Sapienza campuses 5 days per week;
- Students' living arrangements are predominantly linked to their family of origin, whereas staff members typically live in their own family households.

Regarding travel characteristics, the vast majority of both students (86%) and staff (92%) live within the Province of Rome. The main destination campuses are Città Universitaria (approximately 43% of students and 45% of staff), San Pietro in Vincoli (about 10% of students and 9% of staff), and Policlinico Umberto I (around 9% of students and 11% of staff).

Overall, the average travel time for students is 74 minutes from home to university and 81 minutes from university to home, while for staff the corresponding average times are 55 and 65 minutes. These figures alone highlight the low level of sustainability of the mobility system associated with the "Sapienza ecosystem."

With respect to modal choice, based on the predominant mode (i.e., the one on which the most travel time is spent), students mainly rely on public transport, whereas staff primarily use private cars (Figure 3.9).

Overall, the share of public transport users is:

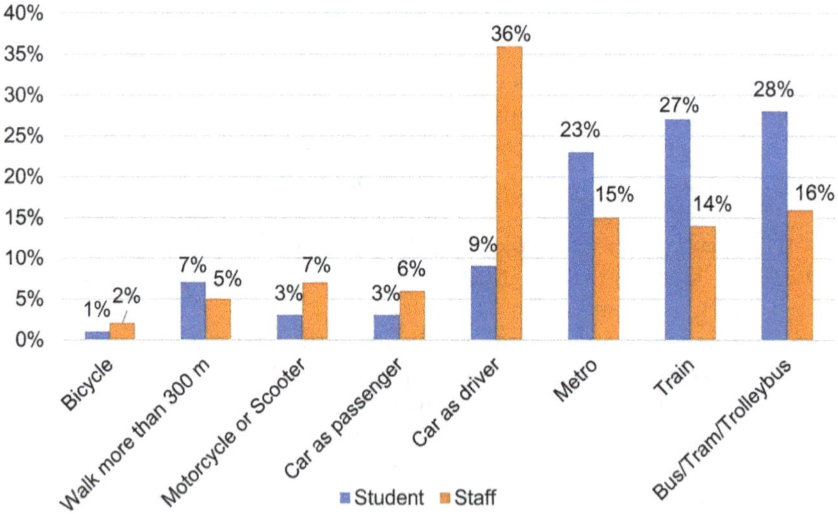

Figure 3.9 Main mode of travel.

Source: Sgarra et al., 2022.

- 78% of students (28% bus/tram/trolleybus, 27% train, and 23% metro);
- 45% of staff (16% bus/tram/trolleybus, 14% train, and 15% metro).
- Travel by private motorized modes accounts for 15% of students and 49% of staff. Only 8% of students and 7% of staff travel either on foot (7% of students and 5% of staff) or by bicycle (1% of students and 2% of staff).

The questionnaire included a particularly insightful section on respondents' perceptions and opinions of their travel experience. Participants were asked to rate, on a scale from 0 (not applicable) to 5 (very important), the level of difficulty perceived during home–university and university–home trips.

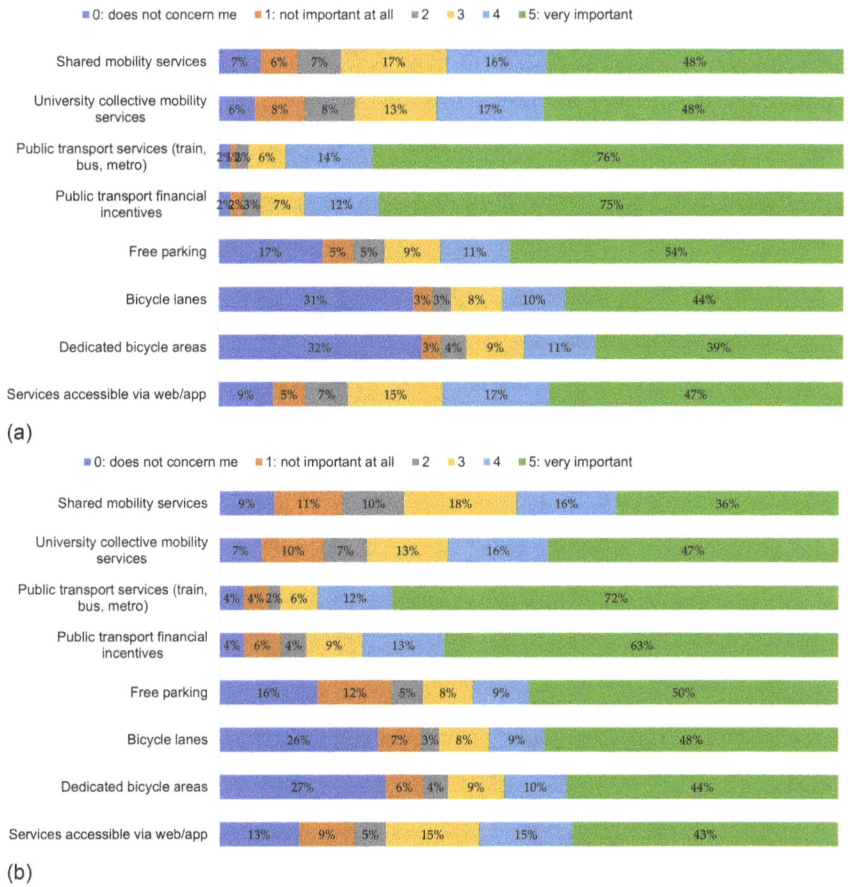

Figure 3.10 Importance assessment of proposed improvement measures for students (a) and staff (b).

Source: CTL, 2020.

Both groups identified difficulties related to public transport services as very important.

For both groups, traffic congestion was the next most significant difficulty, rated as very important by 28% of students and 34% of staff.

Students and staff were also asked to evaluate potential proposals, again using a scale from 0 to 5. Among all proposed measures (Figure 3.10), the highest scores were recorded for:

- Improving public transport services (including bus, rail, and metro): 76% of students and 72% of staff;
- Financial incentives for public transport (including bus, rail, and metro): 75% of students and 63% of staff;
- University collective mobility services: 48% of students and 47% of staff.

Based on the analysis of current mobility conditions, the problems encountered, the stated attitudes of users, and the vision of the University's governance, ten specific objectives were defined, all aimed at improving the overall

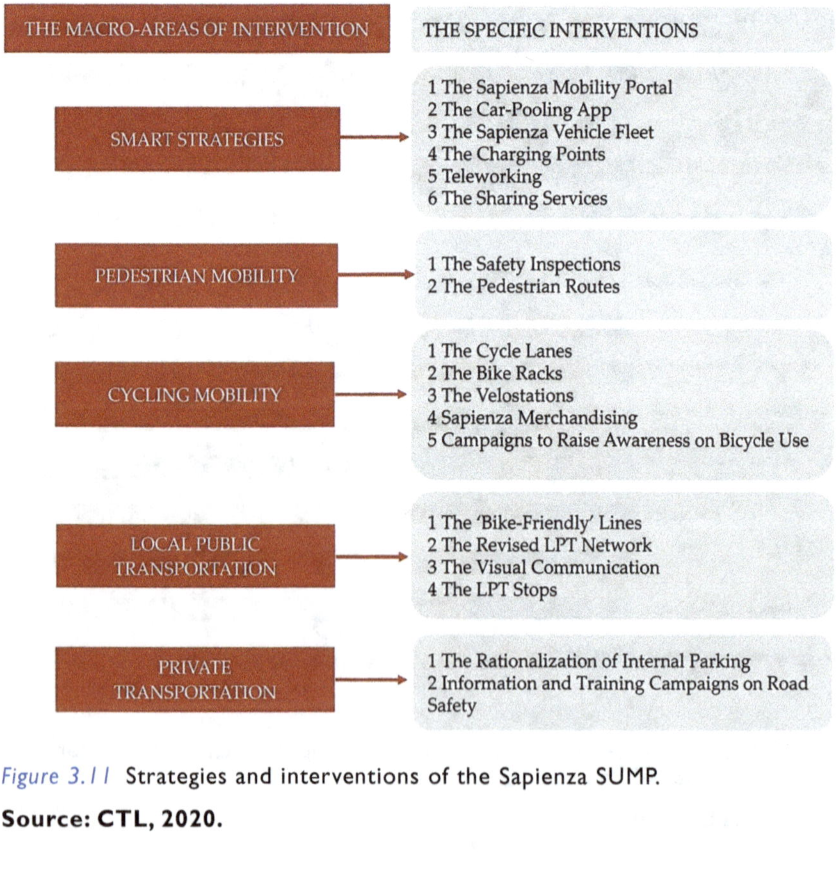

Figure 3.11 Strategies and interventions of the Sapienza SUMP.

Source: CTL, 2020.

sustainability of the university mobility system. This was followed by the identification of strategies, grouped into five Macro-Areas of Intervention (Figure 3.11):

- Smart strategies—6 specific interventions;
- Pedestrian mobility—2 specific interventions;
- Cycling mobility—5 specific interventions;
- Local public transport—4 specific interventions;
- Private transport—2 specific interventions.

NOTE

1 Amsterdam, Athens, Barcelona, Berlin, Bern, Bilbao, Bologna, Brussels, Budapest, Clermont Ferrand, Copenhagen, Dublin, Geneva, Ghent, Glasgow, Graz, Hamburg, Helsinki, Krakow, Lille, Lisbon, London, Lyons, Madrid, Manchester, Marseilles, Milan, Moscow, Munich, Nantes, Newcastle, Oslo, Paris, Prague, Rome, Rotterdam, Seville, Stockholm, Stuttgart, Tallinn, Turin, Valencia, Vienna, Warsaw, Zurich.

REFERENCES

CTL – Centro di Ricerca per il Trasporto e la Logistica, 2009, "Strategie e misure per uno sviluppo sostenibile dei trasporti urbani".

CTL – Centro di Ricerca per il Trasporto e la Logistica, 2020, *Sapienza Sustainable University Mobility Plan*. https://mobilita.uniroma1.it/wp-content/uploads/2024/05/SUMP_SECONDA_FASE.pdf

Federico, A., & Barbabella, A., 2006, *ISFORT - Studio sugli indicatori di valutazione delle politiche per la mobilità urbana sostenibile*. Rome, Italy: ISSI – Istituto Sviluppo Sostenibile Italia, prepared for ISFORT (Istituto Superiore di Formazione e Ricerca per i Trasporti), Osservatorio sulle Politiche per la Mobilità Urbana Sostenibile (OPMUS).

Gudmundsson, H., 2003, "The policy use of environmental indicators-learning from evaluation research", *The Journal of Transdisciplinary Environmental Studies*, 2 (2), 1–12.

ISFORT, 2018, "15° Rapporto sulla mobilità degli italiani." https://www.isfort.it/wp-content/uploads/2019/09/Rapporto_Mobilita_2018.pdf

Jeon, C., & Amekudzi, A., 2005, "Addressing sustainability in transportation systems: Definitions, indicators, and metrics", *Journal of Infrastructure Systems*, 11, 1, 31–50.

Kenworthy, J., & Laube, F., 1996, "Automobile dependence in cities: An international comparison of urban transport and land use patterns with implications for sustainability", *Environmental Impact Assessment Review*, 16, 4–6, July–November, 279–308.

Le BIPE, 2018, *Zero Empty Seats - A Study Into BlaBlaCar's Environmental Contribution*.

Meta, E., Usami, D. S., Azarko, A., Jackson, J., & Verschuur, S., 2020, "ADAS&ME project Deliverable 8.1 – Impact assessment". ADAS&ME (ADAS and Driver Monitoring for Adaptive Support and E-value), Project Deliverable, unpublished/ internal report.

Negri, M., & Bieker, G., 2025. *Life-Cycle Greenhouse Gas Emissions from Passenger Cars in the European Union*. International Council on Clean Transportation.

Newman, P., & Kenworthy, J., 1999, *Sustainability and Cities: Overcoming automobile dependence*, Washington, DC: Island Press. ISBN 978-1559636605.

OECD, 1997, *Towards Sustainable Transportation, Proceedings of an International Conference* (Vancouver, March 1996).

OECD, 2002, *OECD Guidelines towards Environmentally Sustainable Transport*. OECD Publishing, Paris.

Rupprecht Consult (editor), 2019, *Guidelines for Developing and Implementing a Sustainable Urban Mobility Plan*, Second Edition. Rupprecht Consult – Forschung & Beratung GmbH, Cologne, Germany.

Sandén, B., & Wallgren, P., 2017, "System perspectives on electromobility." E-published at: http://www.chalmers.se/en/areas-of-advance/energy/cei/Pages/Systems-Perspectives.aspx

Santos, 2018, "Sustainability and shared mobility models", *Sustainability*, 2018, 10, 3194

Sgarra, V., Meta, E., Saporito, M. R., Persia, L., & Usami, D. S., 2022, "Improving sustainable mobility in university campuses: The case study of Sapienza University", *Transportation Research Procedia*, 60, 108–115.

Ward, J. W., Michalek, J. J., Samaras, C., Azevedo, I. L., Henao, A., Rames, C., & Wenzel, T., 2021, "The impact of Uber and Lyft on vehicle ownership, fuel economy, and transit across U.S. cities", *iScience*. https://doi.org/10.1016/j.isci.2020.101933

World Commission on Environment and Development (WCED), 1987, *Our Common Future*. New York: Oxford University Press.

Chapter 4

Water sustainability: A hydroclimatological perspective (Relevant SDGs: 6, 11, 13, 14, and 15)

Francesco Cioffi

4.1 INTRODUCTION

Water sustainability and climate change are distinct but interrelated issues. Owing to the growing impacts of anthropogenic water use, the environmental, social, and economic sustainability of the water cycle has become a fundamental requirement, irrespective of the occurrence of climate change. Since the onset of the Industrial Revolution, human activities have fundamentally reshaped the global freshwater system through flow regulation, land-use change, and direct water management, altering both the quantity and quality of water resources. These anthropogenic pressures increasingly constrain socioeconomic development, food production, and ecosystem integrity, with converging evidence, such as declining groundwater and surface-water levels and widespread wetland loss, indicating that current patterns of water use are environmentally unsustainable. Nevertheless, climate change compels a further expansion of the concept of water sustainability by challenging existing management frameworks and exposing the limits of traditional, scale-dependent approaches.

Water is an essential element for life as it has evolved on our planet, and the water cycle has played and continues to play a fundamental role in the economic and social development of human communities. This role is evident in the SDGs of the 2030 Agenda, all of which are directly or indirectly related to water.

The problem of water sustainability has traditionally been among the main topics of study in civil, hydraulic, and environmental engineering academic programs. Engineering-driven solutions for the optimal management of water resources among different uses—domestic, industrial, agricultural, and environmental; water purification and wastewater treatment; the estimation of probabilities of extreme hydrological events (floods and droughts); and water-related disaster protection methods—are typical issues addressed within water sustainability.

This traditional approach to water sustainability focuses mainly on the terrestrial branch of the hydrological cycle, at the local basin or regional

DOI: 10.1201/9781003595700-5

scale, namely the part of the cycle related to the transformation of meteorological inputs into surface runoff along river courses, water storage in surface-water bodies such as lakes, infiltration into shallow and deep aquifers, evapotranspiration processes, and soil moisture dynamics. The terrestrial branch of the hydrological cycle concerns natural processes occurring on or immediately below the Earth's surface and is closely linked to the availability of freshwater at the basin or continental scale, thereby directly affecting human water use.

A fundamental assumption of the traditional approach to water sustainability is the possibility of assuming a unidirectional coupling between the atmospheric and terrestrial branches of the hydrological cycle. In this framework, rainfall amounts at different temporal scales are typically represented by appropriate probability density functions whose statistical moments are considered invariant over time.

However, human pressure on the planet, through greenhouse gas emissions and land-use changes such as deforestation or urbanization, alters the temporal and spatial structure of atmospheric circulation, as well as water recycling processes among the different compartments of the climate system, and the transport of water vapor through the atmosphere.

In this new context imposed by anthropogenic pressure, the hydrological cycle must be represented in its entirety, complexity, and temporal variability, accounting for the mutual interactions between its branches and, more generally, for its complex interactions with the climate system.

Consequently, water sustainability requires that boundaries to human actions at the planetary scale be identified as reference points for mitigation strategies aimed at limiting the impact of climate change, while also quantifying the probable changes in the hydrological cycle associated with different mitigation scenarios. These changes constitute the essential knowledge base for identifying and planning adaptation actions at the local scale in specific geographic regions.

This broader concept of water sustainability does not replace the traditional view but rather integrates it into a holistic perspective capable of capturing the mutual coupling among the different branches of the hydrological cycle, the links between local–regional and global-scale phenomena, and the complex interactions between the hydrological cycle and other components of the Earth's climate system.

4.2 WATER IN THE UN 2030 AGENDA

Continuing from the Rio Summit and concluding at the World Summit on Sustainable Development in Johannesburg in 2002, heads of state and government assumed "the collective responsibility to promote and strengthen the three inseparable pillars of sustainable development: environmental protection, economic development, and social inclusion."

These pillars form the foundation of the subsequent 2030 Agenda, signed by 193 UN member states in September 2015, which identifies 17 main goals articulated into 169 sub-goals.

Among the 17 goals of the 2030 Agenda, Goal 6, "Ensure availability and sustainable management of water and sanitation for all," is the most explicitly related to water. However, other goals and their sub-goals also concern the water cycle, such as Goals 11, 13, 14, and 15. These goals and their sub-goals are detailed in Box 4.1.

Although these goals explicitly mention water, as illustrated in Figure 4.1, all of the 2030 Agenda goals are in fact related to water. Such goals fundamentally address four primary objectives of decision-making for water governance: providing drinking water and sanitation services, balancing water supply and demand, maintaining healthy water resources, and adapting to extreme hydrometeorological events (Vardon et al., 2025).

However, the 2030 Agenda expands the concept of water sustainability beyond this traditional view, which posits that the hydro-social water cycle must not compromise the renewability of water resources at the basin, regional, or national level. Instead, it integrates this traditional perspective into a global framework where the water cycle is also viewed at the planetary scale as a regulator of climate, biodiversity, and life in general.

The Paris Agreement in the same year (2015) on limiting greenhouse gas emissions, as well as the United Nations Conference in Sendai on disaster risk reduction (Sendai Framework for Disaster Risk Reduction 2015–2030), highlight that the major challenges posed by climate change at the planetary

BOX 4.1 2030 AGENDA GOALS DIRECTLY RELATED TO THE WATER CYCLE

Goal 6 aims to ensure sustainable water management, both in quantity and quality, to provide adequate drinking water availability to communities and ensure proper wastewater treatment to protect and restore ecosystems. These ecosystems, to a large extent, are also the primary water supply sources.

Goal 6 is structured into the following sub-goals:

6.1 By 2030, achieve universal and equitable access to safe and affordable drinking water for all.

6.2 By 2030, achieve access to adequate and equitable sanitation and hygiene for all, and end open defecation, with special attention to the needs of women, girls, and those in vulnerable situations.

6.3 By 2030, improve water quality by reducing pollution, eliminating dumping, and minimizing hazardous chemicals and materials. This includes halving the proportion of untreated wastewater and significantly increasing recycling and safe reuse globally.

6.4 By 2030, substantially increase water-use efficiency across all sectors and ensure sustainable withdrawals and supply of freshwater to address water scarcity and reduce the number of people suffering from water shortages.

6.5 By 2030, implement integrated water resources management at all levels, including transboundary cooperation where appropriate.

6.6 By 2030, protect and restore water-related ecosystems, including mountains, forests, wetlands, rivers, aquifers, and lakes.

6.7 By 2030, expand international cooperation and capacity-building support to developing countries in water and sanitation-related activities and programs. This includes water harvesting, desalination, water efficiency, wastewater treatment, recycling, and reuse technologies.

6.8 Support and strengthen the participation of local communities in improving water and sanitation management.

However, other goals and sub-goals also address the water cycle, including:

- Goal 11 (*Sustainable Cities and Communities*), which under sub-goal 11.5, aims to significantly reduce the number of deaths and affected people and substantially decrease direct economic losses caused by disasters, including water-related disasters, by 2030, with particular emphasis on protecting poor and vulnerable populations.
- Goal 15 (*Life on Land*), which under sub-goal 15.3, states the need to:
  - "By 2030, combat desertification, restore degraded lands, including those affected by desertification, drought, and floods, and strive to achieve a land degradation-neutral world."
- Goal 13 (*Climate Action*), which under sub-goals 13.1 and 13.2, calls for:
  - Strengthening resilience and adaptive capacity to climate-related hazards and natural disasters in all countries.
  - Integrating climate change measures into national policies, strategies, and planning.
- Goal 14 (*Life Below Water*), which under sub-goal 14.1, aims to prevent and reduce marine pollution, particularly pollution originating from land-based activities, and under sub-goal 14.3, calls for minimizing and addressing ocean acidification, a consequence of increased atmospheric CO_2 levels.

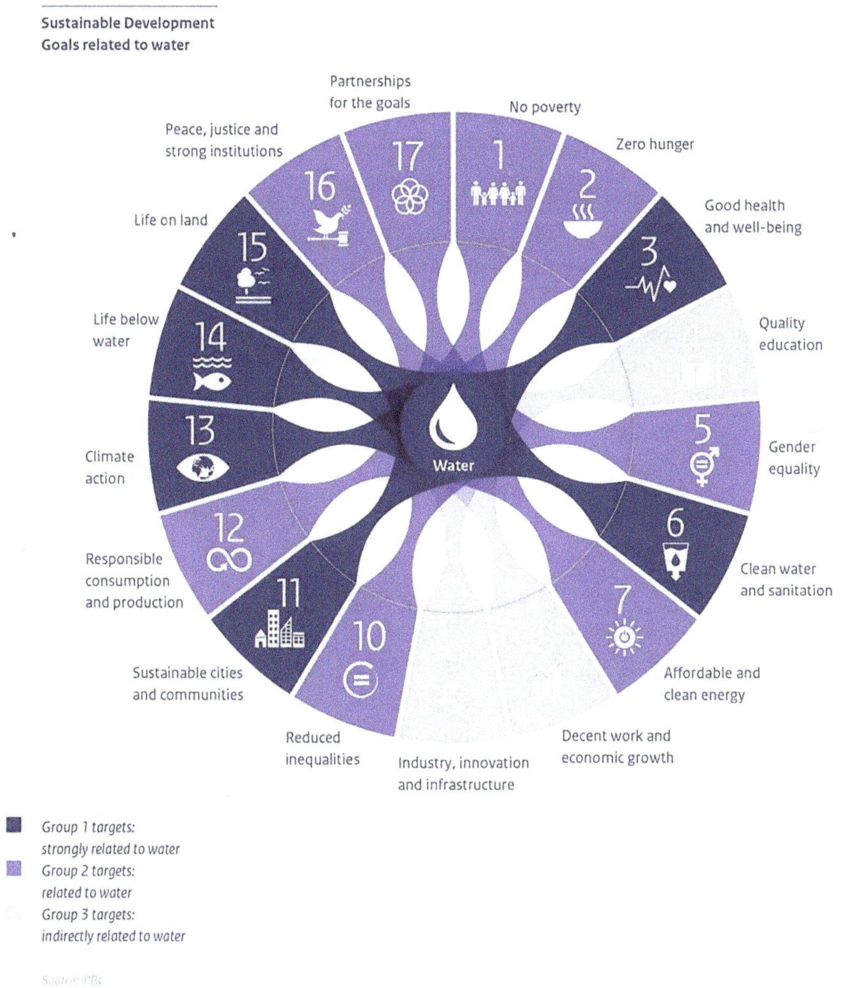

Figure 4.1 Water-related SDGs.

Source: Ward, 2018. https://www.pbl.nl/sites/default/files/downloads/pbl-2018-the-geography-of-future-water-challenges-2920_2.pdf.

scale concern water security. These challenges manifest at local and regional scales as water scarcity, drought, increased flood risk from more intense and frequent extreme hydrological events, and rising temperatures that can alter the quality of aquatic habitats and their biodiversity (UN Office for Disaster Risk Reduction, October 2015).

The 2030 Agenda for Sustainable Development, together with the Sendai Framework for Disaster Risk Reduction (2015–2030), recognizes integrated water resources management (IWRM) as a fundamental operational approach to addressing water-related challenges.

IWRM is a comprehensive framework aimed at managing water resources holistically and sustainably, emphasizing the coordination of policies, institutions, and practices across sectors and regions to ensure equitable and efficient water use. Within the 2030 Agenda, IWRM was adopted as Target 6.5 under Sustainable Development Goal 6 (Nagata et al., 2022). IWRM supports the 2030 Agenda by promoting coordinated management of water, land, and related resources, encouraging participation of all stakeholders, protecting ecosystems, and balancing social, economic, and environmental needs. Since water is fundamental to many other SDGs (e.g., food, energy, climate, health, sustainable cities), progress in IWRM directly contributes to the broader success of the 2030 Agenda.

4.3 WATER AND CLIMATE

4.3.1 The climate system

The Earth's climate system can be described as comprising several interconnected subsystems, including the atmosphere, hydrosphere, cryosphere, biosphere, lithosphere, and anthroposphere (Rial et al., 2004). These subsystems exchange mass, energy, and momentum through the processes shown in Figure 4.2, leading to complex feedback mechanisms that exhibit non-proportional responses, characteristic of nonlinear behavior (Fan et al., 2021).

The nonlinear nature of the system gives rise to unexpected structures, abrupt state transitions, oscillations, and chaotic behavior. Even in deterministic systems, unpredictability can occur. The climate system is both chaotic and complex, composed of many interacting parts that can self-organize, resulting in emergent global-scale phenomena (Ghil & Lucarini, 2020).

These processes occur across a broad range of spatial and temporal scales, from small-scale turbulence (a few meters and seconds) to synoptic-scale weather systems (tens of kilometers and days to weeks) and climate-scale processes (thousands of kilometers and hundreds or thousands of years). At a large scale, the climate system evolves due to solar heating, Earth's rotation, and gravitational forces. Every second, Earth absorbs 1.22×10^{17} joules of energy from the Sun, transported as shortwave electromagnetic radiation according to the Stefan-Boltzmann law:

$$F = \sigma T^{4},$$

where $\sigma = 5.67 \times 10^{-8}\ \mathrm{W\,m^{-2}\,K^{-4}}$ and $T_{\mathrm{Sun}} = 6000\mathrm{K}$, and Earth emits infrared radiation at $T_{\mathrm{Earth}} = 255\mathrm{K}$ (Pierrehumbert, 2011).

Two major atmospheric regions can be distinguished:

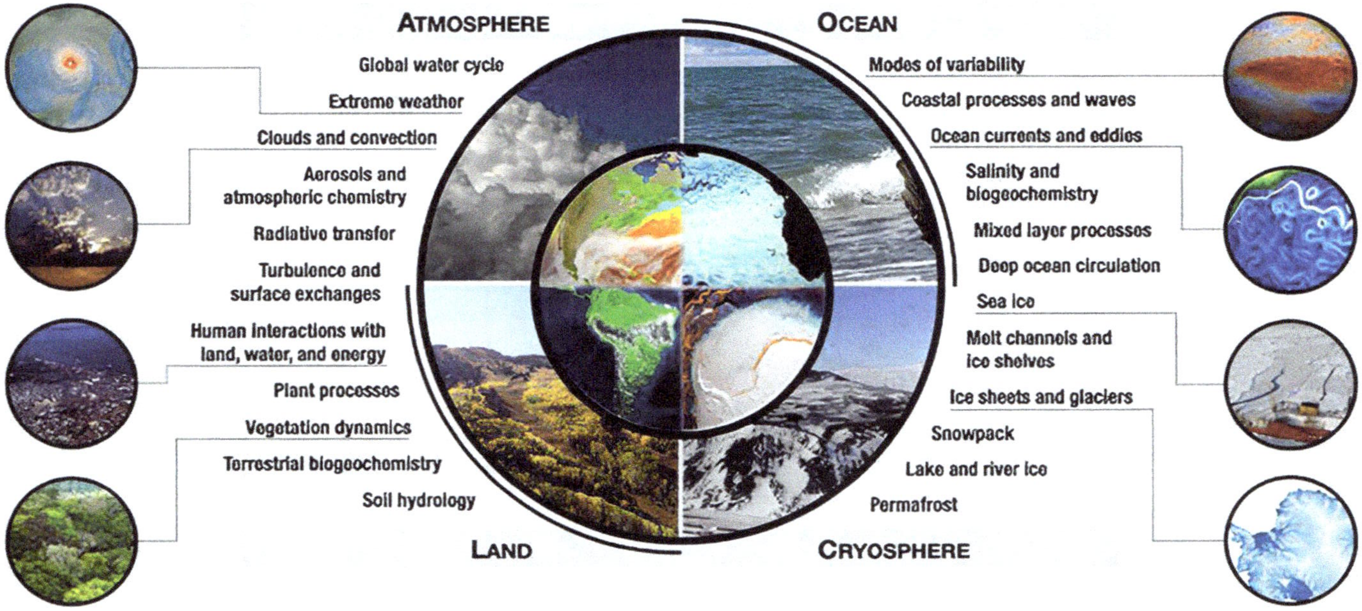

Figure 4.2 The climate system includes many interdependent components and processes.

Source: https://www.energy.gov/science/doe-explainsearth-system-and-climate-models.

- The troposphere, which extends up to about 10,000 meters above Earth's surface, features a temperature decline with altitude.
- The stratosphere, which extends up to 50–60 km, is where temperatures increase due to absorption of solar radiation by ozone.

Within the troposphere, the predominant absorption of energy at the Earth's surface and in the lower atmospheric layers drives multiple processes, such as infrared radiation absorption, phase changes of water, and convective vertical motions, that redistribute energy vertically, as reflected in the atmospheric temperature lapse rate, that is, the rate at which air temperature decreases with increasing altitude.

Although Earth's atmosphere contains various gases, as shown in Figure 4.3, only certain gases absorb infrared radiation: primarily water vapor, carbon dioxide, nitrous oxide, and methane. Nitrogen and oxygen, the most abundant gases in the atmosphere, are essentially transparent to both visible and infrared radiation, while ozone absorbs ultraviolet radiation and a narrow band of infrared radiation around 10 μm (Pacher et al., 2018).

Water absorbs radiant energy in the infrared wavelengths through various rotational and vibrational modes of its molecules, as illustrated in Figure 4.4.

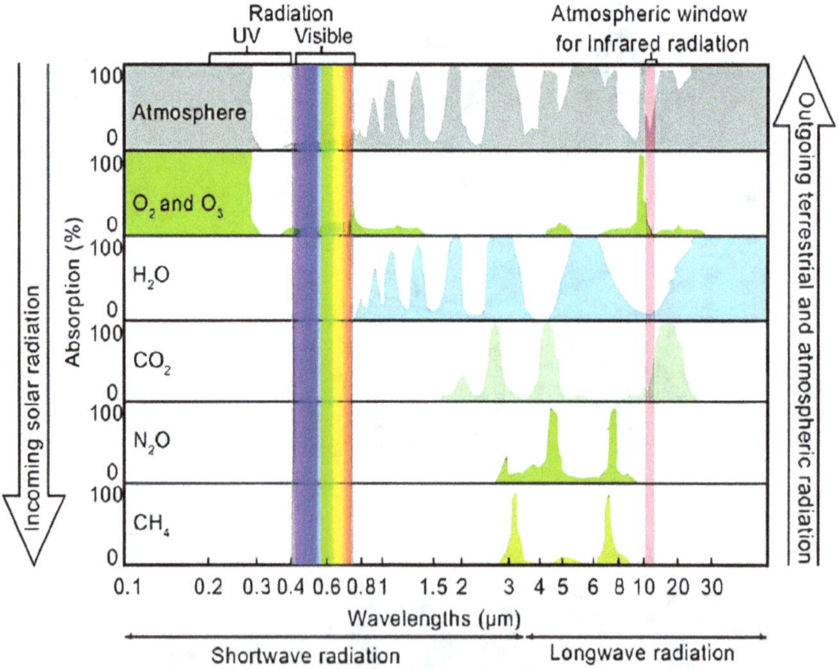

Figure 4.3 Absorption spectra of radiant energy of gases in the atmosphere.

Source: Pacher et al., 2018.

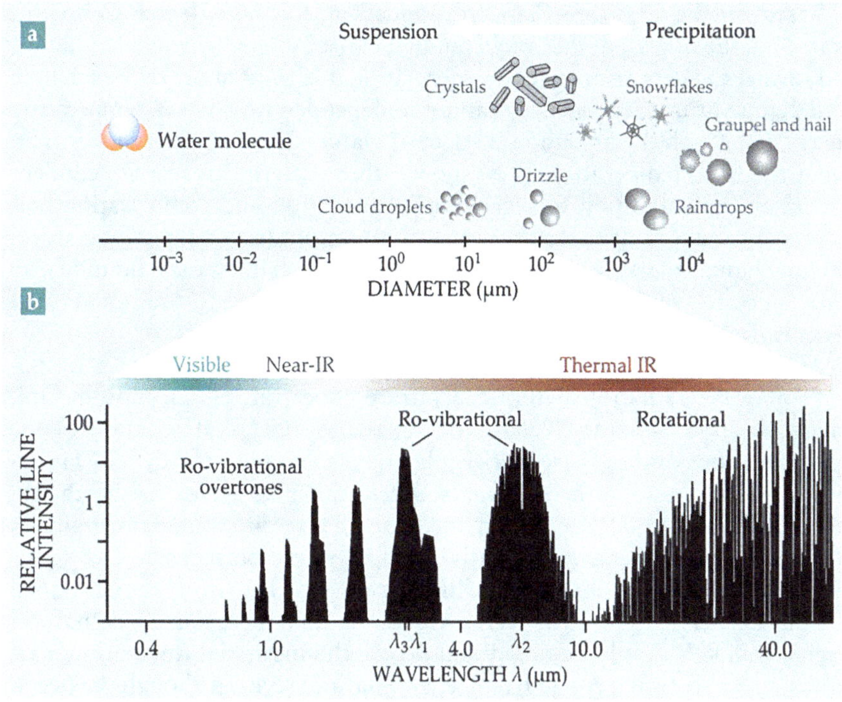

Figure 4.4 Radiative properties of water in the atmosphere.

Source: Stevens & Bony, 2013.

This grants water an essential role in regulating Earth's climate (Stevens & Bony, 2013).

Water vapor and clouds contribute about 60% of the total greenhouse effect (Pacher et al., 2018). However, water vapor resides in the atmosphere for only about nine days before condensing into rain or snow (Van der Ent & Tuinenburg, 2017), and its water vapor saturation pressure responds to temperature according to the Clausius–Clapeyron equation:

$$d\ln e_s = (\beta / T) d\ln T. \quad (4.1)$$

This equation expresses the dependence of water vapor saturation pressure e_s on temperature, effectively constraining atmospheric humidity structure. The factor β is approximately the ratio of the enthalpy of vaporization (the energy required to convert liquid water to gas at constant pressure) to the water vapor constant. Its value of approximately 5400°K is two to three times higher than that of gases such as carbon dioxide or methane. If vapor pressure exceeds (e.g., due to cooling from expansion), condensation occurs, maintaining vapor pressure near saturation (Stevens & Bony, 2013).

Based on the Clausius–Clapeyron equation, for each degree Celsius of warming, the atmosphere's water-holding capacity rises by roughly 7%.

However, excess atmospheric water vapor, due to its short residence time and temperature dependence, cannot independently drive a temperature increase through the greenhouse effect: if water vapor is briefly higher than what the local temperature–pressure–weather pattern can sustain, condensation and precipitation increase until evaporation and condensation balance again. As long as those background conditions (especially temperature) do not change, the atmosphere relaxes back to its characteristic humidity, so extra injected water vapor does not, by itself, keep driving additional long-term warming.

As previously noted, several other atmospheric trace gases absorb outgoing longwave radiation within the infrared spectral region, most prominently carbon dioxide (CO_2) and methane (CH_4). The atmospheric concentrations of these gases are regulated by a complex and dynamic interplay of natural and anthropogenic sources and sinks across the Earth system. Carbon dioxide (CO_2) and methane (CH_4) differ fundamentally in their atmospheric behavior, radiative efficiency, and persistence. The dominant role of CO_2 in long-term climate change arises from the sustained imbalance between anthropogenic emissions and the capacity of natural carbon sinks. It should be noted that before the Industrial Revolution, natural sources were in a rough balance with natural sinks. Although the ocean and terrestrial biosphere currently absorb a substantial fraction of human-induced CO_2 emissions, this uptake remains insufficient to offset total emissions, resulting in continuous atmospheric CO_2 accumulation.

As shown in Figure 4.5, which provides a schematic representation of the overall perturbation of the global carbon cycle caused by anthropogenic activities, averaged globally for the decade 2015–2024, the removal of excess atmospheric CO_2 ultimately depends on slow oceanic and geochemical processes operating on centennial to millennial timescales, implying that a significant fraction of emitted CO_2 persists in the atmosphere for centuries or longer (Archer et al., 2009; IPCC, 2021; Friedlingstein et al., 2025a, 2025b). Kaufhold et al. (2025) showed that, depending on the magnitude of emissions, approximately 75% of anthropogenic CO_2 is removed within 197–1820 years after peak CO_2 concentration is reached, with larger cumulative emissions requiring longer removal times.

In contrast, methane is characterized by a much shorter atmospheric lifetime (~12 years) but a markedly higher radiative efficiency per unit mass. Since pre-industrial times, increased atmospheric CH_4 concentrations have contributed approximately 0.5°C to the observed global mean surface temperature rise, accounting for about 30% of total anthropogenic warming to date. The effective radiative forcing associated with methane is estimated at ~0.54 W m^{-2}, including both direct infrared absorption and indirect effects mediated through tropospheric ozone formation and enhanced stratospheric water vapor (IPCC, 2021; Myhre et al., 2013). These properties are reflected

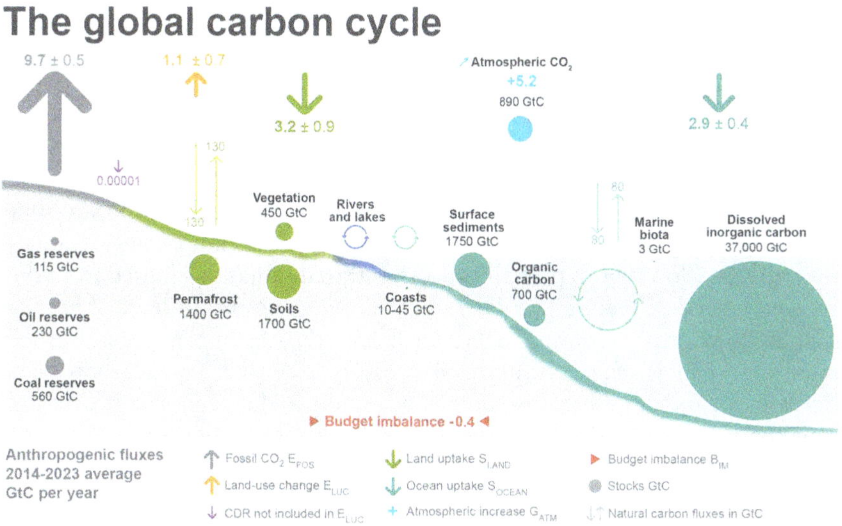

Figure 4.5 The global carbon cycle and schematic representation of the overall perturbation of the global carbon cycle caused by anthropogenic activities, averaged globally for the decade 2015–2024.

Source: Friedlingstein et al., 2025b.

in methane's Global Warming Potential (GWP), which is approximately 27–30 over a 100-year time horizon and 80–85 over a 20-year horizon compared to the heat trapped by the same mass of CO_2, which is assigned a GWP value of 1 (IPCC, 2021).

The increase in CO_2 concentration, as well as in the other greenhouse gases, results in a persistent alteration of Earth's energy balance by reducing outgoing infrared radiation (OIR), that is, the heat energy that Earth emits back into space in the form of infrared (longwave) radiation.

As described by Colman and Soden (2021), increases in CO_2 and other greenhouse gases lead to atmospheric warming and, in accordance with the Clausius–Clapeyron relationship, to higher atmospheric water vapor content, thereby generating strong positive temperature feedback. Furthermore, as shown in Figure 4.3, CO_2 absorbs infrared radiation at wavelengths not absorbed by water vapor, partially narrowing the "infrared window" through which energy can escape into space, thereby making a non-negligible contribution to the positive feedback. This positive feedback is partially offset by negative feedback associated with a decrease in the lapse rate, that is, the rate at which atmospheric temperature decreases with increasing altitude. While quantifying the relative magnitude of these climate system feedbacks to increasing greenhouse gases is challenging, rough estimates suggest that the water vapor feedback roughly doubles the response to greenhouse gas forcing compared to a scenario without feedbacks (Pacher et al., 2018).

Colman and Soden (2021) estimate the combined strength of these feedbacks at 1.25 ± 0.15 W m^{-2} K^{-1}.

In addition to water vapor and lapse rate feedbacks, the climate system exhibits several other mechanisms that influence its response to greenhouse gas forcing. Positive feedbacks include the ice–albedo feedback, whereby reductions in snow and ice cover decrease surface reflectivity and increase solar absorption, and the cloud feedback, whose net effect is likely positive but remains a major source of uncertainty due to competing shortwave and longwave effects. Clouds both partially absorb infrared radiation emitted from the surface and reflect incoming solar radiation, reducing the amount of energy reaching the Earth's surface. Further feedback arises from changes in surface properties, such as vegetation and soil moisture, as well as from carbon cycle processes, including reduced oceanic CO_2 uptake and greenhouse gas release from thawing permafrost. Together, these feedbacks, acting on a range of timescales, affect the planetary energy balance and modulate both the magnitude and spatial distribution of climate change (Trenberth et al., 2009).

On the surface of the planet, most solar radiation is absorbed at low latitudes, with varying degrees of absorption between continental areas and oceans, generating non-uniform warming on a planetary scale. This results in thermal gradients between the equator and the poles, as well as between land and oceans. Together with the action of gravity and the Coriolis force due to Earth's rotation, these gradients contribute to the complex structure of atmospheric circulation shown in Figure 4.6.

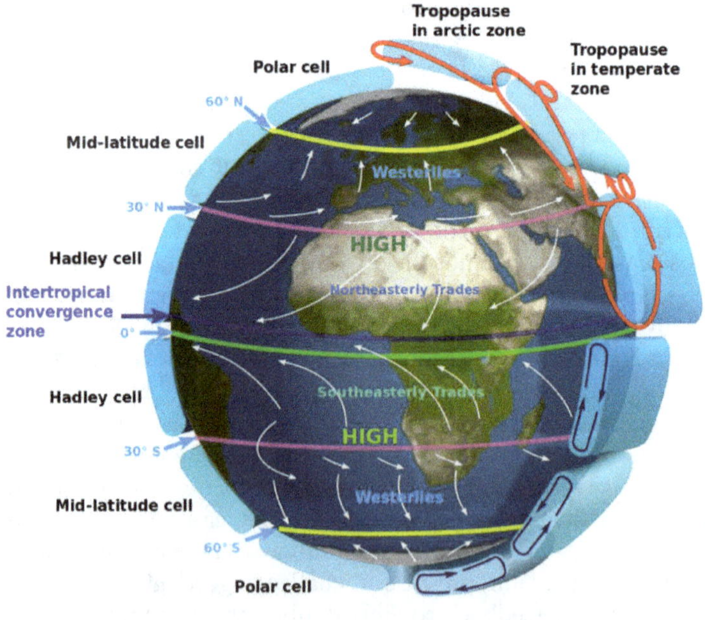

Figure 4.6 The structure of atmospheric circulation.

Source: Fan et al., 2021.

This global circulation, driven by trade winds in tropical regions, has a well-defined organization in each hemisphere: three major convective cells in the meridional planes are associated with five winds along the parallels. These include the weak equatorial easterly current near the equator at low altitudes, two westerly winds in each hemisphere, the polar jet stream at a latitude of approximately ±60° at high altitudes, and the subtropical jet stream, which is slower than the former and located at latitudes around ±30°, also at high altitudes. The three main convective cells are the Hadley cell, the Ferrel cell, and the polar cell. The Coriolis force prevents convective cells from remaining in the meridional planes, imposing a helical atmospheric circulation. More details on atmospheric circulation are provided in Box 4.2.

Winds, the Earth's rotation, gravity, and differences in temperature and salinity between different parts of the oceans all create a very complex oceanic circulation system. This circulation helps regulate the planet's climate and biodiversity by storing and moving heat, nutrients, and carbon. Its changes are influenced by interactions among the Pacific, Atlantic, and Indian Oceans, as well as by the dynamic link between the ocean and the atmosphere (Wang et al., 2020). Coherent circulation patterns, such as the thermohaline circulation (also known as the Gulf Stream) (Broecker, 2010) and the patterns connected to El Niño, both have a global impact on climate (Box 4.3).

BOX 4.2 ATMOSPHERIC CIRCULATION

In polar regions, a convective circulation is driven by the descent of cold, dry, and dense air from the upper layers of the troposphere. Due to the stronger influence of Coriolis forces, convective circulation near the poles remains confined between them and the parallels at ±60°. Between the latitudes of the Hadley cell (±30°) and the polar cell (±60°), another convective cell, known as the Ferrel cell, is established. Between the Hadley and Ferrel cells, air descends to the surface after cooling and drying at high altitudes, creating the desert belt between the tropics and temperate regions. In contrast, the ascending zones, located near the equator in one case and between the polar and Ferrel cells in the other, experience heavy precipitation.

In each hemisphere, below the tropopause (at an altitude of approximately 8–10 km above mean sea level), westerly winds circulate around the planet at high altitudes on both sides of the Ferrel cells, oscillating around a mean latitude. These west-to-east winds are generated by the Coriolis force in narrow zones characterized by large temperature differences between warm air to the south and cold air to the north, as well as significant pressure differences. The polar jet stream is very fast (with speeds ranging between 100 and 300 km h^{-1}) and highly turbulent, while the tropical jet stream is slower (ranging from 50 to 100 km h^{-1}) and more stable. These air masses transport and redistribute both the heat transferred from continents and the moisture produced by ocean evaporation.

BOX 4.3 OCEANIC CIRCULATION STRUCTURES: THERMOHALINE CIRCULATION AND EL NIÑO

Both circulation structures, the thermohaline circulation and El Niño, influence the climate on a planetary scale. The thermohaline circulation, shown in Figure 4.7, is responsible for redistributing heat across all oceans. In the Atlantic Ocean, the Gulf Stream transports warm, saline waters northward, heating the atmosphere at high latitudes. The cooling of these waters at such latitudes

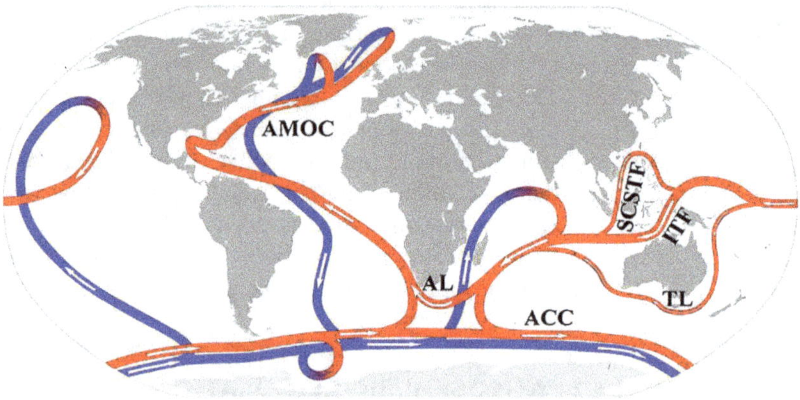

Figure 4.7 The structure of the thermohaline circulation (Great Conveyor Belt). The thermohaline circulation, an invisible conveyor belt that transports the warm, light, and salty waters of the Gulf Stream toward the North Atlantic, starts on the surface of the North Atlantic Ocean and is driven by the Atlantic Meridional Overturning Circulation (AMOC). The red color indicates warm surface water, and the blue color indicates cold deep water. ITF indicates the Indonesian Throughflow; SCSTF is the South China Sea Throughflow; AL and TL are the Agulhas Leakage and the Tasman Leakage, respectively; and ACC is the Antarctic Circumpolar Current.

Source: Wang, 2019.

increases their density, causing the current to sink and return southward in deep ocean layers, eventually reaching the Indian and Pacific Oceans.

The structure of the thermohaline circulation ultimately depends on temperature and salinity gradients between the equatorial and polar regions of the Atlantic Ocean. The current configuration is dominated by the temperature gradient between equatorial and polar waters. As demonstrated by Stommel (1961) and Rahmstorf (1995), the thermohaline current has a bistable equilibrium. Under changing conditions, such as those induced by global warming, it may reverse its course, leading to a drastic shift in heat, nutrient, and carbon fluxes between the equator and high latitudes.

El Niño is a phenomenon that exemplifies the complex interaction between the ocean and the atmosphere. This phenomenon involves the formation, at intervals of two to five years, of a positive temperature anomaly in the Eastern Equatorial Pacific Ocean, where, under normal conditions, a negative anomaly is present due to vertical water recirculation induced by winds blowing westward in the equatorial Pacific region (Cane, 2005). The increase in surface-water temperatures in the Eastern Equatorial Pacific weakens these winds, causing the migration of the warm surface-water region from the Western Pacific to the east. This process further weakens the easterly winds, amplifies the temperature anomaly in the Eastern Equatorial Pacific, and reverses vertical oceanic recirculation.

El Niño is a phenomenon driven and amplified by nonlinear positive feedback mechanisms dependent on ocean–atmosphere coupling.

As shown in Figure 4.8, El Niño influences the global climate, with an impact that depends on the duration and intensity of the phenomenon. The thermohaline current and El Niño highlight, even more than other phenomena, the complex, chaotic, and nonlinear nature of the Earth system (Lucarini, 2013).

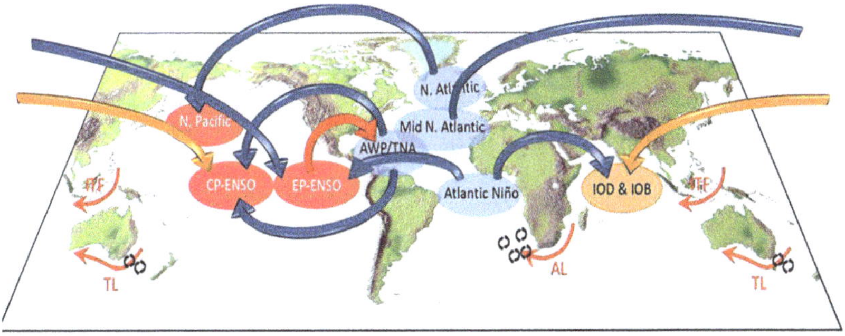

Figure 4.8 Schematic diagram summarizing the interaction processes of the three oceans. The thick arrows indicate the direction of interaction, which can be unidirectional or bidirectional. The bidirectional curved arrow means that two oceans can influence each other: for example, ENSO in the Eastern Pacific (EP) and Central Pacific (CP) can influence the Indian Ocean Dipole IOD/IOB, which in turn can influence ENSO events. ITF indicates the Indonesian Throughflow, and AL and TL are the Agulhas Leakage and the Tasman Leakage, respectively, while AWP and TNA represent the Atlantic Warm Pool and the Tropical North Atlantic, respectively.

Source: Wang, 2019.

4.3.2 Water cycle

The concept of the water cycle, or hydrological cycle, has ancient philosophical and descriptive origins. It emerged with the civilizations of great rivers in ancient Asia and the Middle East, which were based on the development of agriculture and the ability to create large-scale irrigation systems and protective measures against floods and droughts. Descriptions of the hydrological cycle can also be found in Aristotle (384–322 B.C.) and Marcus Vitruvius (second half of the 1st century B.C.).

Although these descriptions were not far from actual phenomena, a scientific approach to representing the hydrological cycle only began to develop in the 15th century. This involved identifying the cause-and-effect relationships among its different components and quantifying them. Figures such as Leonardo da Vinci (1452–1519), Bernard Palissy (1510–1589), Pierre Perrault (1608–1680), Edme Mariotte (1620–1684), and Edmond Halley (1656–1742) contributed to this process. However, it was only in 1802 that John Dalton formulated a complete and accurate representation of the hydrological cycle, supported by experimental observations (Dalton, 1802; Nace, 1975). This representation is shown in Figure 4.9 and is essentially the same as that found in school textbooks and manuals (Abbott et al., 2019). For practical purposes, the hydrological cycle is considered a closed system involving the three different phases of water.

Typically, two branches of the hydrological cycle are distinguished: the terrestrial branch, which includes continental processes referred to as the freshwater that flows on or below the land surface (blue water), stored in sources such as rivers, lakes, aquifers, and reservoirs, and evapotranspiration and soil moisture; and the atmospheric branch, which describes the processes of moisture and energy redistribution associated with the

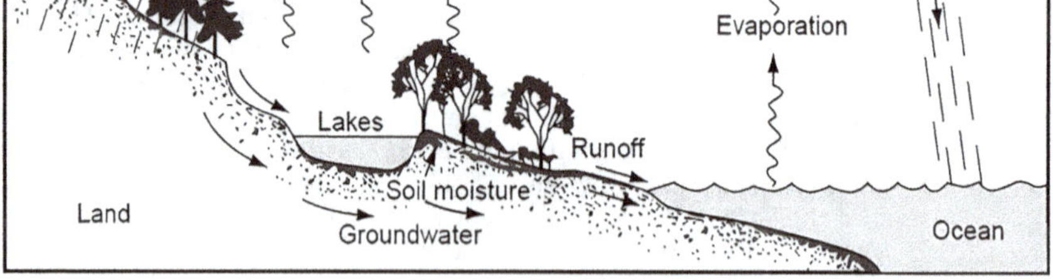

Figure 4.9 The water cycle or hydrological cycle.

transport, evaporation, and condensation of water in the atmosphere (Shelton, 2009) and precipitation (green water). The representation, formalization, and modeling of these processes are essential for the sustainable management of water resources and for protecting against disasters caused by extreme hydrological events.

The terrestrial and atmospheric branches are interconnected through evaporation and evapotranspiration processes, which transfer water from the terrestrial to the atmospheric branch of the hydrological cycle, and through liquid or snow precipitation, which returns water to the terrestrial branch. As described in previous sections, the atmospheric branch of the hydrological cycle is integrated into atmospheric circulation, whose structure at different scales determines the transport of moisture between various regions of the planet, cloud formation processes, and thermodynamic changes associated with phase transitions of water.

Although, for practical reasons, the two branches have been studied separately within the disciplines of hydrology, meteorology, and climatology, the hydrological cycle must be examined as a whole.

In reality, the hydrological cycle is not a perfect cycle. Its patterns are shaped by interactions with other cycles, quasi-cycles, and random events within the Earth system. These include changes in Earth's orbit, solar cycles, both regular and irregular oscillations in the ocean–land–atmosphere climate system, and the activities of living organisms that rely on water in all its forms (Duffy, 2017).

From a planetary perspective, water in its various states regulates both climate and biodiversity through what Gleeson et al. (2020a) define as "hydroclimatic, hydroecological, storage, and transport water functions," which are regulated by the following phenomena:

- the exchange of water between the atmosphere, land surface, soil, ice, snow, oceans, and underground aquifers influences Earth's climate by mediating energy, carbon, and water balance;
- as stated in Section 4.3.1, water vapor is the most powerful greenhouse gas due to its infrared absorption spectrum, high heat capacity, and abundance in the atmosphere. Water vapor forms clouds that reflect solar radiation and absorb infrared radiation, affecting the planet's energy balance. The cycles of evaporation and condensation are the primary mechanisms by which heat is redistributed from the equator to the poles;
- soil moisture, surface water, and glaciers directly or indirectly influence Earth's surface albedo and, therefore, the radiative balance. Additionally, soil moisture and surface waters act as sources or sinks of carbon through processes such as photosynthesis, soil oxygenation, and carbon transport and accumulation (see the chapter on Ecology and Ecosystem Systems).

- Surface water, atmospheric water, soil moisture, groundwater, and glaciers are natural reservoirs of freshwater that interact with Earth's system components: land, oceans, and the atmosphere.

Table 4.1 illustrates in more detail the main functions of water within the Earth system, including their key processes, primary reservoirs, main drivers, feedbacks, and roles in the Earth system. According to Gleeson et al. (2020a), these functions are defined with reference to water's role at the planetary scale, extending the functions traditionally defined in hydrological disciplines, which mainly focus on describing the terrestrial branch of the hydrological cycle at regional or basin scales.

4.4 WATER SUSTAINABILITY

As highlighted in Section 4.2 and in Figure 4.1, all of the United Nations' 2030 Agenda goals are directly or indirectly related to water. The availability of freshwater in adequate quantity and quality is an essential condition for ensuring economic and social development as well as environmental protection, the three inseparable elements that define the concept of sustainability.

Approximately 3,800 km^3 $year^{-1}$ of renewable freshwater (from rivers, lakes, and groundwater) is used by human communities. This represents less than 10% of the freshwater available on Earth. Additionally, another 30% comes directly from precipitation. Although this may seem like a low percentage, in reality, water is available in unequal quantities and quality across different regions of the planet, and many areas experience water stress (Oki & Kanae, 2006).

As evidenced in the 2023 UN World Water Development Report, the 26% of the world's population, approximately two billion people, do not have access to safe drinking water, and 3.6 billion lack access to safe sanitation services. Over 400 million people still practice open defecation. This lack of sanitation, which includes unimproved facilities like open pits or buckets, is a major public health crisis, contributing to disease and death, particularly among children under five. It also has high economic costs and hinders development and education, particularly for women and girls.

He et al. (2021) estimated for 2016 that 359 million (12.5%) and 573 million (20.0%) of urban residents experienced perennial and seasonal water scarcity, respectively. Water scarcity in urban areas is expected to worsen, as projections indicate a doubling of the affected population from 930 million in 2016 to between 1.7 and 2.4 billion people by 2050.

In recent decades, water-related disasters, including droughts, riverine and flash floods, landslides, and tropical cyclones, have increased in both frequency and intensity. Van Dijk et al. (2025) have shown that extremely dry months have become increasingly common, with 38% more record-dry months in 2024 than during the baseline period. Furthermore, daily

Table 4.1 Water functions and their attributes

Domain	*Water function*	*Key processes*	*Primary reservoirs/ components*	*Main drivers*	*Feedbacks & Earth system role*
Hydroclimate (Atmospheric water cycle)	Climate regulation through atmospheric water transport, radiative forcing, and energy transfer	Evaporation, condensation, precipitation, atmospheric moisture transport, cloud formation, atmospheric rivers, radiative water vapor properties	Atmospheric water vapor, cloud systems, storm systems	Solar radiation, atmospheric circulation, ocean surface temperature, aerosols	Controls global precipitation patterns; regulates energy balance; influences storm intensity and climate variability
Cryosphere storage function	Long-term freshwater storage and climate buffering	Snow accumulation, glacier formation, ice melt, sublimation, and permafrost thaw	Ice sheets, glaciers, seasonal snowpack, frozen soils	Temperature gradients, seasonal solar forcing, and climate variability	Alters sea level; affects planetary albedo; modulates freshwater supply
Surface-water storage function	Intermediate water storage and redistribution	River flow regulation, lake storage, wetland water retention, floodplain storage	Oceans, lakes, rivers, wetlands, reservoirs	Precipitation, runoff, snowmelt, river basin hydrology	Regulates flood control; supports biodiversity; transports sediments and nutrients
Subsurface storage function	Long-term buffering and delayed water release	Infiltration, groundwater recharge, percolation, aquifer storage, capillary rise	Soil moisture, vadose zone, groundwater aquifers, deep geological water	Soil properties, precipitation infiltration, vegetation uptake, and geological structure	Maintains drought resilience; supports baseflow in rivers; sustains ecosystems
Hydrologic transport function	Movement of water across the Earth system	River discharge, groundwater flow, ocean circulation, atmospheric advection, snowmelt runoff	River networks, subsurface flow systems, ocean currents, and atmospheric moisture pathways	Gravity, pressure gradients, temperature gradients, and topography	Connects reservoirs globally; redistributes heat, freshwater, and nutrients

(*Continued*)

Table 4.1 (Continued)

Hydro-ecological function	Ecosystem regulation through water availability and cycling	Root water uptake, transpiration, canopy interception, wetland filtration, soil moisture cycling	Vegetation systems, soils, wetlands, and riparian ecosystems	Climate conditions, plant physiology, land cover, nutrient availability	Controls ecosystem productivity; regulates carbon and nutrient cycles; influences precipitation recycling
Biogeochemical coupling function	Transport and transformation of nutrients and carbon via water	Nutrient leaching, dissolved organic carbon transport, sediment transport, wetland biogeochemistry	Soil organic matter, wetlands, river sediments, ocean biogeochemical pools	Water flow intensity, microbial activity, temperature, and vegetation productivity	Links the water cycle to the carbon and nutrient cycles; influences climate feedbacks
Ocean–atmosphere coupling function	Exchange of moisture and heat between ocean and atmosphere	Sea surface evaporation, heat exchange, storm generation, and thermohaline circulation	Ocean surface waters, deep ocean circulation, and sea surface temperature gradients	Solar radiation, wind patterns, ocean salinity, and temperature gradients	Drives global climate systems; influences rainfall distribution and extreme weather
Soil moisture–climate interaction function	Regulates land–atmosphere energy and moisture exchange	Evapotranspiration, infiltration, soil evaporation, and root uptake	Soil moisture layers, root zone water	Vegetation cover, rainfall patterns, and land surface properties	Controls drought development; influences convective precipitation formation
Anthropogenic water modification function	Human alteration of natural hydroclimate processes	Dam regulation, irrigation withdrawals, groundwater pumping, urban runoff, and land-use change	Reservoirs, irrigation systems, urban drainage systems, and water diversion infrastructure	Population growth, agriculture, industry, and land management practices	Alters natural flow regimes; modifies climate feedbacks; impacts ecosystem stability

precipitation extremes were 52% more common in 2024 than during 1995–2005, with record-breaking daily rainfall events reported in West Africa, Europe, and Asia. In 2024, water-related disasters caused more than 8,700 fatalities, displaced 40 million people, and resulted in global economic losses exceeding US$550 billion; the true figures are likely higher due to incomplete data and unreported events.

It should also be emphasized that unsustainable water resource management practices inherited from the past have exacerbated these impacts. River flow regimes have been irreversibly altered by dams or diverted, in some cases, leading to the complete drying of rivers and their downstream lakes. Groundwater reserves have been depleted, and many water bodies, including estuaries, coastal areas, and even oceans, have become polluted, resulting in the degradation of habitats and ecosystems. In coastal areas, excessive groundwater extraction has contributed to land subsidence and saltwater intrusion into aquifers. Flood-protection structures have often been constructed without consideration of their ecological impacts or their effects on the hydrological balance; measures intended to address floods or droughts can sometimes have unintended negative consequences on the risk of the opposite hazard (Ward et al., 2020).

Many of these unsustainable practices were driven by short-term economic goals, without adequate consideration of their long-term environmental, social, and economic sustainability or the associated health impacts.

Sustainable water use requires maintaining a balance between the natural water cycle (precipitation, surface runoff, and storage in surface or underground water bodies) and water withdrawals for domestic, agricultural, and industrial use. This balance must ensure ecosystem integrity and the renewability of both water quantity and quality for current and future generations. Focusing on the terrestrial branch of the hydrological cycle, that includes all the processes by which water moves, is stored, and is transformed on land, involving soils, vegetation, surface waters, and groundwater, sustainable water management must account for the variability of hydrological flows affecting surface and deep-water bodies at multiple spatial and temporal scales. This variability ranges from individual river basins or hydrogeological domains to regional scales and even cross-border scales, as in the case of the Jordan River, the Nile, and the Danube. Rainfall and river flow patterns are highly variable throughout the year, with seasonal fluctuations between wet and dry periods, as well as multi-year oscillations due to natural climate variability. This means that, in most cases, water resources are not available when and where human communities need them. There are different solutions to address this challenge, including building dams and flow control structures (temporal relocation of water), watershed restoration through reforestation, water transfer between basins (spatial relocation), desalination, recycling and reuse, and pollution control (Boberg, 2005; Gleick, 2003a). The last group of solutions focuses on reducing water consumption, including limiting the use of potable water by collecting rainwater for non-drinking purposes, reducing water losses in supply infrastructure, recycling water within the same production cycle, and reusing wastewater to recharge groundwater. Sustainable water management from a demand-side perspective involves optimal water allocation among its various social, economic, and environmental uses. These uses often compete and are measured in incommensurable units that are not always easily quantifiable in monetary terms. For example, a conflict exists between hydropower generation and the preservation of river habitats downstream of a dam (Cioffi & Gallerano, 2012). In this case, the objectives of maximizing electricity production and ensuring the survival of aquatic species are non-directly comparable because the ecological health of a river habitat cannot be directly expressed in economic terms, unlike electricity production. In such cases, multi-objective optimization techniques must be used to balance competing goals and help decision-makers identify the best possible compromise. Similar challenges arise when balancing social inclusion goals with productive uses of water resources. To address such challenges, IWRM, a coordinated approach to managing water, land, and related resources to maximize social and economic benefits while ensuring environmental sustainability, was formally introduced as a global policy framework in 1992, following the Dublin International Conference on Water and the Environment and the Rio Earth Summit (Agenda 21), and it was later incorporated into the 2030 Agenda for Sustainable Development. There is extensive literature on IWRM (Liu et al., 2017). Focusing on the terrestrial branch of the hydrological cycle, the traditional approach in water management assumes that precipitation and surface runoff (derived from the atmospheric branch of the water cycle) follow natural variability patterns and can be represented using statistical hydrology. Generally, the design guidelines and standards developed for hydrologic and hydraulic infrastructure and management of water resources systems are based on the stationarity assumption, that implicates the time-invariance of statistical properties of a time series of climate variables. Based on this approach, certain design parameters, such as the flood discharge for a given return period, are used as reference points for designing hydraulic engineering structures. However, as discussed in Section 4.5.5, in the context of global warming, the concept of hydrological cycle sustainability requires a profound revision, as, for instance, the assumption of statistical stationarity that historically guided hydrological analysis, infrastructure design, and water resources management becoming increasingly invalid. In fact, the assumption of stationarity may lead to over- or under-design when climate variable time series are nonstationary.

As will be discussed in the following sections, observational and modeling evidence point to an intensification of the global hydrological cycle. This intensification is characterized by an increased capacity of the atmosphere to retain water vapor, more intense precipitation, and greater spatial-temporal variability, accompanied by an increase in the frequency and intensity of extreme events such as droughts and floods, with strongly heterogeneous responses at the regional scale. Such intensification does not necessarily translate into greater water resource availability, as increases in potential evapotranspiration, rapid runoff, and anthropogenic land-use changes tend to reduce effective groundwater recharge, shorten water residence times, and weaken the natural storage function of the hydrological system. Within this framework, hydrological sustainability can no longer be assessed exclusively in terms of long-term average balances at the local scale of one or more interconnected basins, but must be reinterpreted in light of the non-stationarity of hydro-climatic processes, that is, the temporal variation of the statistical distributions of precipitation, streamflow, and recharge, driven by both climate change and anthropogenic pressures on land-use changes, as deforestation (Ranzi et al., 2025).

4.5 CLIMATE CHANGE

4.5.1 Climate change: Evidence

Over the last three centuries, the anthropogenic impact resulting from population growth and exploitation of the planet's resources has become the dominant factor altering the climatic and ecological balances of the Earth system, to the point of suggesting the definition of a new geological epoch: the Anthropocene (Crutzen, 2002). During this time, the human population has increased tenfold, along with urbanization and global forest loss due to deforestation through permanent land-use change for commodity production. This has led to an unprecedented loss of biodiversity (see Chapter on ecology and ecosystem services) and a significant alteration of the climatic balance of the biosphere.

The release of greenhouse gases, such as carbon dioxide, methane, nitrogen oxides, and halogenated compounds, has significantly altered the Earth's energy balance with respect to the last 10,000 years, during which Earth remained within the Holocene Stability domain (Rockstrom et al., 2009a). In response to this increase in greenhouse gases, a substantial rise in global temperatures has been observed.

As reported by the Global Carbon Project (Friedlingstein et al., 2025b) and other observers, 2025 also set a new record for atmospheric carbon dioxide concentrations due to the continued accumulation of CO_2 from human activities. The annual amount of carbon dioxide emitted in 2025 was 1.1% higher than in 2024, reaching a new all-time high. Measurements of air trapped in ice cores, together with ongoing analyses of modern air samples, indicate that global atmospheric CO_2 concentrations have risen from about 280 ppm in the pre-industrial atmosphere to over 430 ppm in December 2025 (https://climate.nasa.gov/vital-signs/global-temperature/).

Total cumulative carbon emissions from 1850 to 2023 amount to approximately 715 GtC, of which about 490 ± 25 GtC derive from fossil fuels and industry, and 225 ± 65 GtC (802 ± 220 $GtCO_2$) from land-use change (Friedlingstein et al., 2025b). In the 1960s, atmospheric carbon dioxide increased by about 0.8 ppm per year on average. That growth rate doubled to an average of 1.6 ppm per year in the 1980s and 1.5 ppm per year in the 1990s. In the last decade (2015–2024), the annual increase has accelerated to 2.6 ppm per year (https://www.climate.gov/news-features/understanding-climate/climate-change-atmospheric-carbon-dioxide#:~:text=According%20to%20observations%20and%20analysis,produced%20greenhouse%20gases%20since%201990). Regnier et al. (2013) estimated that approximately half of CO_2 anthropogenic emissions remain in the atmosphere, with the remainder absorbed by the oceans and the terrestrial biosphere. IPCC AR6 WG showed that CO_2 accounts for about two-thirds of the total anthropogenic radiative forcing associated with long-lived greenhouse gases (2.076 W m^{-2} out of 3.140 W m^{-2}), excluding methane, ozone, aerosols, and clouds.

The current atmospheric concentration of methane (CH_4), approximately 1800 ppb, is unprecedented over at least the last 800,000 years (Loulergue et al., 2008), having increased by about 160% since pre-industrial times. Observations from global atmospheric monitoring networks, mainly NOAA and AGAGE, show that global methane concentrations rose rapidly until the late 1990s, stabilized in the early 2000s, and then began increasing again from 2006 onward (Folke et al., 2021). Although uncertainties in the global methane budget remain, the stabilization period is thought to be mainly due to fluctuations in natural sources and sinks, whereas the continued increase observed since 2007 is largely attributable to human activities. As highlighted in Section 4.3.1, given methane's high GWP, reducing anthropogenic methane emissions is essential to limiting global warming (Turner et al., 2019).

The increase in anthropogenic greenhouse gases is directly responsible for the observed global temperature rise since 1850. Comparative analyses using global circulation models have shown that neither natural climate variability, volcanic eruptions, nor variations in solar radiation can explain this temperature increase (Myhre et al., 2013). Furthermore, paleoclimatic reconstructions covering the recent few centuries to the 20th century and instrumental data show that greenhouse gases began to cause warming since the beginning of industrialization, with trends attributable to greenhouse gases evident by 1900 in proxy-based temperature reconstructions. However, other forcings have caused substantial deviations from this emerging greenhouse warming trend: volcanic eruptions have caused strong cooling following a period of unusually heavy activity, such as in the early 19th century, or warming during periods of low activity, such as in the early-to-mid-20th century (Hegerl et al., 2019).

Figure 4.10 shows the global surface temperature trends reconstructed by the Met Office Hadley Centre/Climatic Research Unit global dataset (Morice

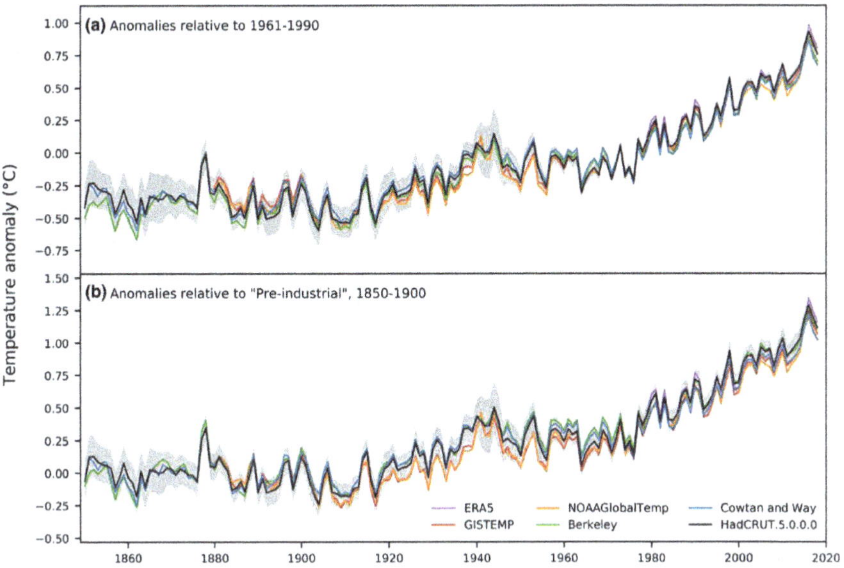

Figure 4.10 Comparison of the series of annual global mean temperature anomalies (°C), reconstructed by the Met Office Hadley Centre/Climatic Research Unit Global, relative to two baselines, 1961–1990 (a) and 1850–1900 (b), taken as representative of pre-industrial conditions. Gray: HadCRUT5 analysis ensemble mean; red: ERA5; orange: GISTEMP; green: NOAA Global Temp; blue: Berkeley Earth; light blue: Cowtan and Way; gray shading: 95% confidence interval on the HadCRUT5 analysis.

Source: Morice et al., 2021.

et al., 2021). The figure clearly illustrates the sharp acceleration in temperature rise since 1990 (Allen et al., 2018).

In Berkeley Earth's analysis (https://berkeleyearth.org/global-temperature-report-for-2025) the global mean temperature in 2025 is estimated to have been 1.44 ± 0.09°C (2.60 ± 0.17°F) above the average temperature of the period 1850–1900, which is often used as a pre-industrial baseline for global temperature targets. Moreover, land areas generally exhibit about twice as much warming as the oceans. Compared to the 1850–1900 average, the land mean temperature in 2025 increased by 2.03 ± 0.17°C (3.66 ± 0.32°F), while the ocean surface temperature, excluding sea-ice regions, increased by 1.03 ± 0.05°C (1.86 ± 0.09°F). Most of this warming has occurred since 1970. Earth's average surface temperature in 2024 was the warmest on record since instrumental observations began in 1880 (source: NASA/GISS). This temperature increase exhibits a high global spatial-temporal coherence: 98% of the planet has recorded higher temperatures than at any time in the last 2,000 years. Such global and synchronous warming has never occurred in pre-industrial eras, where natural climate drivers were insufficient to cause global temperature variations over multi-decadal and century-long timescales (Neukom et al., 2019).

While global warming affects the entire planet, it is not uniform. In the Northern Hemisphere, between 60°N and 90°N latitude, temperatures have increased by approximately 3°C compared to 1960–1990 (https://climate.copernicus.eu/esotc/2024/arctic-temperature). In contrast, between 0°N and 30°N, the increase has been only about 0.8°C. Differences in temperature between continental areas and oceans, as well as between tropical and mid-to-high latitudes, have altered zonal and meridional temperature gradients, which, along with gravity and Earth's rotation, are the main drivers of atmospheric circulation.

The 20th century recorded a decreasing trend in global temperature gradients, which has contributed to changes in atmospheric circulation structures, including the strength of the Hadley cell, jet streams, storm tracks, and planetary waves (Dong et al., 2025; Karamperidou et al., 2012; Keel, 2025).

These atmospheric circulation changes affect moisture transport mechanisms across different regions of the planet, impacting precipitation patterns. Additionally, higher temperatures increase atmospheric moisture content due to stronger evaporation processes in accordance with the Clausius–Clapeyron equation.

Since 1950, a significant increase in the severity, duration, and frequency of droughts has been observed in various regions, including the Mediterranean, Southern Australia, Central and South America, West and Southern Africa, and East Asia (Chiang et al., 2021; Spinoni et al., 2019). Furthermore, as shown by Yuan et al. (2023) on the basis of estimates of soil moisture from three global reanalyzes from 1951 to 2014, anthropogenic climate change is driving the transition to flash droughts, that is, droughts lasting for 0–40 days with a rapid onset speed due to the intensifying rainfall deficit and increasing evapotranspiration which dries the soil quickly and creates ideal conditions for heat waves.

Additionally, in some regions, the 20th century has seen an increase in the intensity and frequency of extreme precipitation events, influenced by large-scale atmospheric circulation (Cioffi et al., 2015; Donat et al., 2016). These extreme events have intensified even in drought-prone areas, amplifying the variability of the hydrological cycle. As a result of rising ocean surface temperatures, studies indicate that between 1982 and 2009, tropical cyclones, especially in the Atlantic, became more intense and destructive (Bhatia et al., 2022).

One of the clearest signs of anthropogenic climate forcing is the melting of glaciers, a phenomenon of particular concern due to its critical role in Earth's climatic balance. Recent studies, including Hugonnet et al. (2021), have shown that between 2000 and 2019, glaciers worldwide lost an average of 267 ± 16 gigatons of ice per year, contributing to 21 ± 3% of the observed sea-level rise during the same period. The study also found an acceleration in ice mass loss of 48 ± 16 gigatons per decade, accounting for 6–9% of the acceleration in sea-level rise. More recently Dussaillant et al. (2025) have shown that since 1976, Earth's glaciers have lost 9179 ± 621 Gt (187 ± 20 Gt per year) of water, contributing 25.3 ± 1.7 mm (0.5 ± 0.2 mm per year) to the global mean sea-level rise. About 41 % (~ 10 mm) of this loss occurred in the last decade, with 6 % (~ 1.5 mm) occurring in 2023 alone, the record-breaking year for glacier mass loss.

4.5.2 Non-linearity, feedback mechanisms, critical thresholds, and tipping points in the climate system

At the beginning of Section 4.3.1, it was emphasized that the climate system is a chaotic and complex system, composed of different subsystems that are nonlinear, dissipative, forced, and far from equilibrium. As highlighted by Ghil and Lucarini (2020), under certain circumstances, even small perturbations of natural or anthropogenic origin can trigger a variety of phenomena, such as positive feedback loops (amplified and disproportionate responses to disturbances), phase transitions from one system state to another, which may be reversible or irreversible and may exhibit hysteresis effects, bifurcations, where transitions can occur gradually or abruptly, but the system's future evolution depends on exceeding a critical threshold of the disturbance.

In climate science, these phenomena are collectively referred to as "tipping points." Lenton et al. (2008) formally defined a "tipping element" as a subsystem of the Earth system that has at least a subcontinental spatial scale (~1,000 km) and, under certain conditions, can undergo abrupt changes between qualitatively different states due to small perturbations.

In this formal definition, the tipping point corresponds to the critical threshold of a forcing factor or a system parameter beyond which the system's future state is qualitatively altered; that is, self-amplifying feedback in response to the forcing qualitatively reorganizes the system. The geological history of the Earth provides numerous examples of nonlinear state changes in the climate system, including the rapid desertification of the Sahara, which occurred approximately 6,000 years ago, and Dansgaard–Oeschger (D/O) oscillations, which took place during the last glaciation (~60,000 years ago) (Ganopolski & Rahmstorf, 2001).

In the context of current climate change, scientific interest is focused on specific tipping elements and their corresponding critical thresholds, which could trigger state transitions in the climate system with impacts on social and economic systems within a century-scale timeframe, making them politically relevant. However, some scholars argue for looking beyond this temporal scale and for considering the long-term survival of human civilization by aligning its trajectory with Earth's deep dynamics. Within this framework, Gerya et al. (2026) proposed, in the context of climate change, a hierarchical approach based on four characteristic bio-geodynamical and societal timescales ranging from 100 years to millions of years. In this approach, short-term goals (<100 years) are aimed at mitigation and adaptation to the climate effects of short-term anthropogenic changes related to global technological, agricultural, and population changes, while long-term goals aim to ensure that the future Earth will still have a stable climate and a rich ecology upon which our distant descendants will depend.

Table 4.2 and Figure 4.11 list the main climate tipping elements, along with their associated critical thresholds (Lenton et al., 2008). For a detailed description of each tipping element, refer to Box 4.4.

As indicated in Table 4.2, qualitative state changes in climate tipping elements are ultimately linked to exceeding global temperature thresholds, typically between 3 and 5°C above pre-industrial levels. However, recent report by the Intergovernmental Panel on Climate Change (IPCC) suggest that these critical thresholds may be significantly lower, between 1 and 2°C (Hoegh-Guldberg et al., 2018)).

Another critical concern in future climate dynamics is the possibility of a domino effect between different tipping elements (Lenton et al., 2019; Wunderling et al., 2024), where Arctic ice loss amplifies regional warming, Greenland ice melt increases freshwater input into the North Atlantic, which slows down the AMOC, a weakened AMOC destabilizes the African Monsoon, leading to drought in the Sahel. Further AMOC slowdown increases drought frequency in the Amazon, disrupts the Asian Monsoon system, warms the Southern Ocean, and accelerates Antarctic ice melt.

If such a chain reaction were to occur, Earth's climate could shift into a completely different state, potentially incompatible with current human social and economic structures (Steffen et al., 2018).

Deutloff et al. (2025) developed a simple modeling framework to explore the probability of triggering climate tipping elements, showing that if anthropogenic emissions continue along the current trajectory, there is a high probability of instantaneous triggering on average across all 16 tipping elements listed in Table 4.1.

However, current climate models still struggle to precisely define critical thresholds and to fully simulate these nonlinear processes involved in tipping elements, and some scientists argue that, due to the limited capability of the simplified models to represent the highly complex reality (Armstrong McKay, 2024), the probability of such tipping point cascades remains uncertain.

Nevertheless, given the potential social and economic impacts of such processes, their irreversibility, and the paleoclimatic evidence indicating that abrupt climate changes have occurred in the past, often leading to mass extinctions (Glikson, 2018), it would be prudent to adopt the precautionary principle to mitigate the risk of crossing these tipping points by limiting greenhouse gas emissions.

Table 4.2 Category, threshold, timescale, and impact of the TEs

Category		*Proposed climate tipping element (tipping point)*	*Threshold (°C)*			*Timescale (years)*			*Max. Impact (GtC)*
			Est.	*Min*	*Max*	*Est.*	*Min*	*Max*	
Global core tipping elements	PFTP	Boreal Permafrost (collapse)	4.0	3.0	6.0	50	10	300	125–250
	GRIS	Greenland Ice Sheet (collapse)	1.5	0.8	3.0	10k	1k	15k	
	WAIS	West Antarctic Ice Sheet (collapse)	1.5	1.0	3.0	2k	500	13k	
	EAIS	East Antarctic Ice Sheet (collapse)	7.5	5.0	10.0	?	10k	?	
	EASB	East Antarctic Subglacial Basin (collapse)	3.0	2.0	6.0	2k	500	10k	
	AWSI	Arctic Winter Sea Ice (collapse)	6.3	4.5	8.7	20	10	100	
	AMAZ	Amazon Rainforest (dieback)	3.5	2.0	6.0	100	50	200	Partial: 30 Total: 75
	AMOC	Atlantic M. O. Circulation (collapse)	4.0	1.4	8.0	50	15	300	
	LABC	Labrador- Irminger Seas SPG Convection (collapse)	1.8	1.1	3.8	10	5	50	

(*Continued*)

Table 4.2 (Continued)

Category		Proposed climate tipping element (tipping point)	Threshold (°C)			Timescale (years)			Max. Impact (GtC)
			Est.	Min	Max	Est.	Min	Max	
Regional impact tipping elements	PFAT	Boreal Permafrost (abrupt thaw)	1.5	1.0	2.3	200	100	300	Adds 25-75% to gradual thaw: = 5–15 GtC per°C @2100; 12.5–37.5 GtC per°C @2300, up to max. 60-195 GtC
	BARI	Barents Sea Ice (abrupt loss)	1.6	1.5	1.7	25	?	?	
	GLCR	Mountain Glaciers (loss)	2.0	1.5	3.0	200	50	1k	
	REEF	Low-latitude Coral Reefs (die-off)	1.5	1.0	2.0	10	–	–	
	SAHL	Sahel and W. African Monsoon (greening)	2.8	2.0	3.5	50	10	500	
	BORF	Boreal Forest (southern dieback)	4.0	1.4	5.0	100	50	?	52 GtC/–0.18°C
	TUND	Boreal Forest (northern expansion)	4.0	1.5	7.2	100	40	?	–6 GtC/+0.14°C
Threshold-free nonlinear Feedbacks	PFGT	Boreal Permafrost (gradual thaw)	1.5	1.0	2.4	300	100	300<	20 GtC per°C @2100; 50 GtC per°C @2300, up to max 260 GtC

Source: Reproduced from Armstrong McKay et al. (2022).

Note: Colors of the second column on the left represent the Earth system domain of the tipping point (blue: cryosphere; green: biosphere; orange: ocean/atmosphere). All other colors represent confidence, with green being high confidence, yellow medium confidence, and red low confidence.

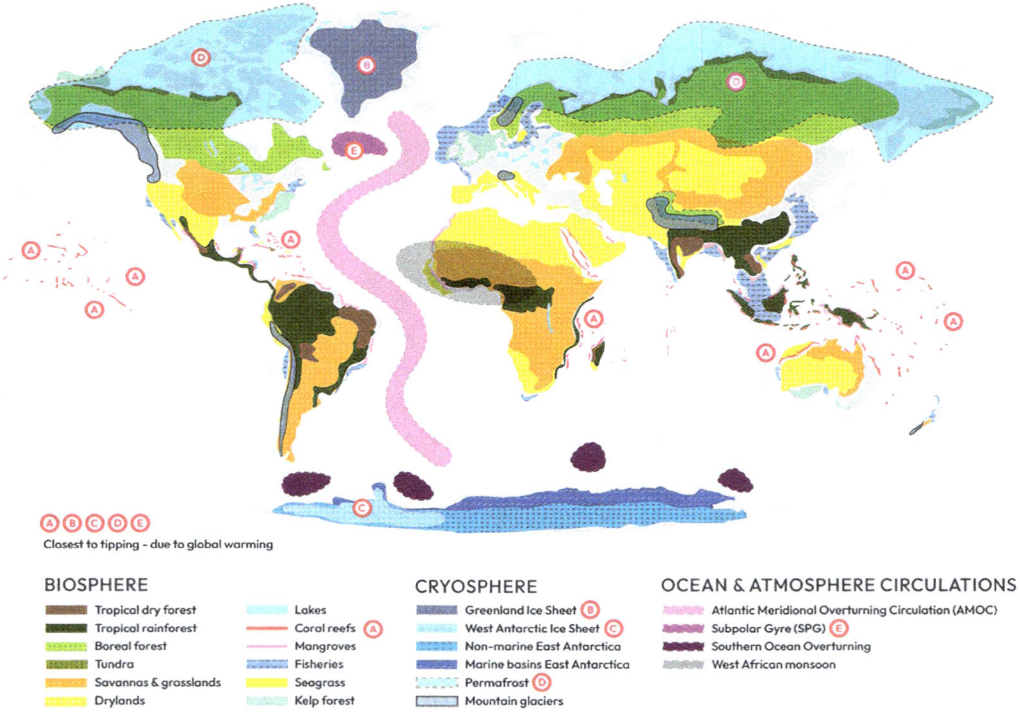

Figure 4.11 Map of potential tipping elements across the biosphere, cryosphere, and circulation in the ocean and atmosphere.

Source: https://research-portal.uu.nl/ws/files/274148919/GTP-2025-FULL-REPORT-FINAL-4-111025.pdf)

BOX 4.4 TIPPING POINTS OF THE CLIMATE SYSTEM

Here, we note that tipping points, shown in Table 4.1, can be categorized based on the specific threshold parameter that governs their state transition. For example temperature acts as a controlling parameter in ice-melting processes in the Arctic, Greenland, Antarctica, and permafrost regions: rising temperatures lead to a decrease in ice volume and surface area, which in turn reduces albedo (reflectivity), causing further warming, increases sea levels and, due to the releases of methane (CH_4) and carbon dioxide (CO_2) from thawing permafrost, further amplifying atmospheric warming. Hydrological cycle changes, that are ultimately caused by alterations in planetary temperature gradients, affect large-scale atmospheric and oceanic circulation. Such changes, particularly shifts in precipitation patterns and increased evaporation rates, influence tipping elements such as the Atlantic Meridional Overturning Circulation (AMOC) (which includes the Gulf Stream) cessation, the Amazon Rainforest dieback, and the Sahel and West Africa greening.

The Atlantic Meridional Overturning Circulation (AMOC) is affected by both increased freshwater input at high latitudes (due to melting of glaciers and poleward shift of the NH storm tracks), and water warming of North Hemisphere polar regions which reduce the water density gradients between the equator and the polar ocean regions, weakening the AMOC, whose collapse in turn can result in profound changes in North Hemisphere circulation (Orbe et al., 2023). In this extreme case, these changes could lead to significant global climate shifts, like those observed in past geological periods.

Another example of positive feedback is related to the Amazon Rainforest. Most of the Amazon's precipitation comes from recycled evapotranspiration within the forest itself. Deforestation and land-use changes reduce the amount of recycled water, leading to drier conditions that hinder forest regeneration. Drier conditions also increase the likelihood of wildfires, which release additional CO_2 into the atmosphere, further accelerating forest loss (Golding & Betts, 2008). It has also been observed that drought conditions in the Amazon are often associated with El Niño events, which reduce rainfall over the basin. El Niño events could become more intense and persistent with global temperature increases of 3–4°C. A detailed description of climate tipping points can be found in Lenton et al. (2008) and Armstrong McKay et al. (2024).

4.5.3 Climate models

The only tools currently available for making quantitative projections about the future evolution of Earth's climate are climate models. A climate model is a simplified representation of reality, formalized through a set of mathematical equations based on the principles of physics, chemistry, and thermodynamics. For some phenomena, primarily biological in nature, empirical laws are also incorporated.

As shown in Figure 4.12, over time, models of varying levels of complexity have been developed.

In constructing a climate model, there is always a trade-off between the need to include as many processes and mechanisms as possible to provide a realistic description of the climate system and the need to have a manageable and sufficiently simple tool that allows scientists to infer cause-and-effect relationships between model forcings and the evolution of the system's state.

At one extreme are process models, that is, zero-dimensional models (or models with only a few degrees of freedom), specifically designed to investigate a single phenomenon using a limited set of variables. Examples include radiative balance models and box models (Houghton et al., 2014). In the 1950s, simple energy balance models were used to examine key processes such as the ice–albedo feedback. Between 1970 and 1980, one-dimensional radiative–convection equilibrium models helped scientists understand and represent the role of water vapor in the atmospheric energy balance.

Box models developed in the 1960s include, for example, Stommel's (1961) model of thermohaline circulation and the Lorenz (1984) model (Lorenz, 1963, 1984, 1996), which captures key features of mid-latitude atmospheric dynamics. These models highlighted the existence of multiple equilibrium states in ocean circulation and the chaotic nature of atmospheric flow.

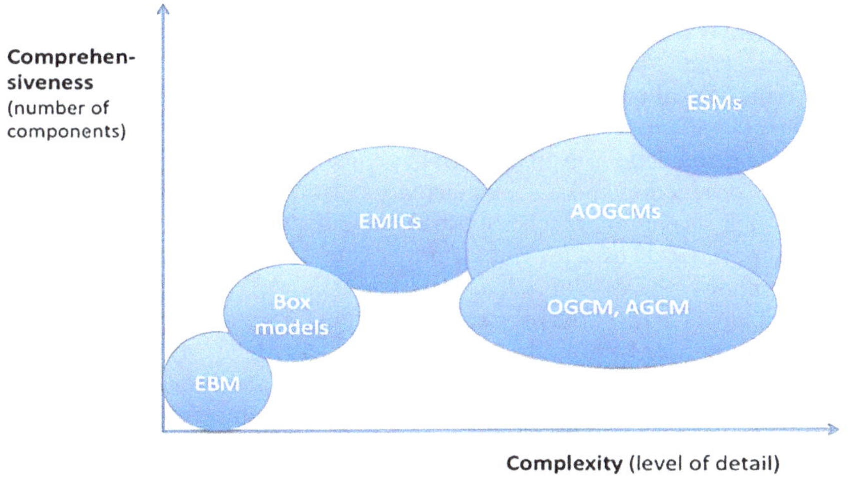

Figure 4.12 Climate models of different complexity and Comprehensiveness (Obe, 2018)

At the other extreme, global climate models (GCMs) aim to represent most of the physical, chemical, and biological processes occurring within the climate system. Given the system's complexity, researchers typically adopt a reductionist approach. In practice, this involves decomposing the climate system into its main components, modeling each separately, and then coupling them to account for interactions among them. There are different ways of applying this reductionist method (Provenzale, 2014).

One of the most common is the subsystem or "spheres" approach, which is adopted in most GCMs. Within this framework, the atmosphere, hydrosphere, biosphere, lithosphere, and anthroposphere are treated as distinct modules in which the different physical, chemical, and biological processes were represented and formalized separately. Historically, models of these components were often developed independently to investigate specific aspects of the climate system, but they are now combined within integrated modeling frameworks.

Another method is the "biogeochemical cycle approach," which treats the major cycles, such as the hydrological, carbon, nitrogen, phosphorus, and sulfur cycles, as interacting systems. Rooted in chemistry and ecology, this perspective emphasizes the central role of the carbon, nitrogen, and hydrological cycles in regulating the functioning of the climate system.

A third strategy, known as the tipping element approach, instead focuses on individual feedback mechanisms capable of triggering large-scale climate changes. These include the ice–albedo feedback, interactions between temperature and water vapor, vegetation–precipitation dynamics, and the complex interplay among clouds, aerosols, and temperature. Between highly simplified process models and fully coupled GCMs, there is also a class of models of intermediate complexity. A particularly important category within this middle ground is that of Integrated Assessment Models (IAMs). Unlike GCMs, IAMs combine the basic physics, thermodynamics, and biology of the Earth system with simplified representations of economic, demographic, and technological processes. These models are especially useful for evaluating climate mitigation policies, as they enable researchers to explore scenarios involving reductions in greenhouse gas emissions, technological advances in energy production and use, and changes in land management.

Although several approaches exist for investigating climate phenomena, today's most widely used for climate projections are the GCMs based on the subsystem or "spheres" approach.

The main climatic processes represented and formalized in a coupled ocean–atmosphere-land model circulation model (AOGCM) are shown in Figure 4.13. These processes are described through a system of differential equations derived from the conservation of mass, momentum, and energy, applied to a rotating sphere in which the fluid medium, air or water, is subject to gravity. The governing equations also include thermodynamic terms to account for radiative transfer and the latent heat released during phase

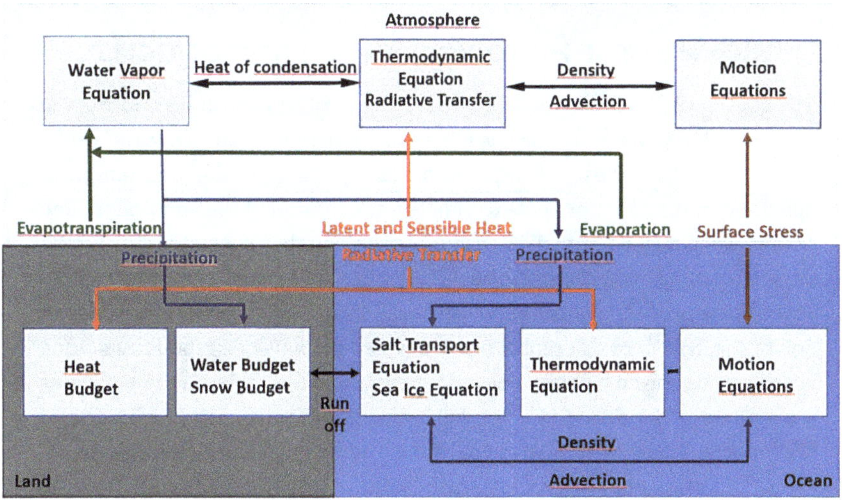

Figure 4.13 The main processes represented in a General Circulation Model (GCM).

Adapted from Manabe & Stouffer, 1999.

changes of water, as well as the energy response to greenhouse gas forcing. Biological processes are also included, for example, via parameterizations of vegetation evapotranspiration, which significantly influences land–atmosphere interactions and the surface energy and water budgets.

Thus, GCMs simulate the large-scale circulation of the atmosphere and oceans, as well as the associated distributions of temperature and salinity, phase changes of water, radiative energy exchanges as modulated by atmospheric composition, albedo, and the thermal inertia of the oceans and continental surfaces. In some cases, these models also represent biogeochemical processes associated with the carbon and nutrient cycles, together with their interactions with vegetation. In such cases, they are referred to as Earth System Models (ESMs). A brief overview of the historical evolution of climate model development is provided in Box 4.5.

The growing awareness of the urgency and complexity of the climate problem, as emphasized by the scientific community, has fostered the development of international research programs in which simulations produced by different GCMs are systematically compared under standardized experimental protocols, including consistent initial conditions and external forcings.

Among these initiatives, the Coupled Model Intercomparison Project (CMIP), launched in 1995, is widely regarded as a cornerstone of climate science (Touzé-Peiffer et al., 2020). The main objectives of CMIP are to:

BOX 4.5 THE HISTORICAL EVOLUTION OF GCMS

The core of GCMs is constituted by the Navier–Stokes equations (Constantin & Foias, 1988). These are partial differential equations that allow the calculation of the quantities (velocity and pressure) characterizing the motion of fluids and thus make it possible to simulate the motion of the atmosphere and oceans. Since they do not admit analytical solutions, they necessarily have to be integrated using numerical methods. These consist of transforming, through an appropriate process of spatial and temporal discretization on discrete grids of points, the system of differential equations into systems of algebraic equations, which can be solved by electronic computation. The variables in the equations are defined at the points of a computational grid, so the spatial resolution of the model is determined by the size of the grid. In simple terms, motion structures, such as vortices or processes that occur at spatial scales smaller than the grid size, cannot be directly simulated, but their effects on the resolved scales must be quantified using more or less simplified sub-grid models.

The Navier–Stokes equations also form the basis of numerical weather forecasting models over time scales of a few days, whereas climate models operate over decades and centuries. In fact, climate models and weather forecasting models share a common origin, which can be traced back to October 1903, at the conference of the Stockholm Physical Society, where the Swedish physicist Vilhelm Bjerknes proposed a rational method for weather forecasting based on a system of seven equations in primitive variables capable of representing atmospheric motion, derived from the Navier–Stokes equations (Bjerknes, 1904). In 1922, Lewis Fry Richardson attempted to numerically integrate Bjerknes' system of equations, laying the groundwork for the development of numerical integration techniques for the equations governing atmospheric motion (Richardson, 1922).

It was only after the Second World War, with the development of electronic computing, that it became possible to build atmospheric circulation models based on the numerical resolution of the equations of motion (Lynch, 2008; Weart, 2010). Pioneering figures in this field included John von Neumann and Jule G. Charney, who in the 1950s built the first weather forecasting model for the United States; Norman A. Phillips, who in 1956 at Princeton proposed the first model to include heat in the lower layers of the atmosphere among the system's forcings (Phillips, 1956); Syukuro Manabe, Joseph Smagorinsky, and Robert F. Stricker, who in 1965 incorporated equations representing the hydrological cycle into an atmospheric circulation model (Manabe et al., 1965); and Edward Norton Lorenz, who at MIT from the 1960s onward highlighted the chaotic nature of atmospheric motion and the limits it imposes on the predictability of weather states (Lorenz, 1963, 1984, 1996).

Between 1950 and 1970, computing power increased by several orders of magnitude, more sophisticated and efficient motion-resolution schemes were developed, and there emerged a growing interest in explicitly climatic aspects. For example, in 1975, Syukuro Manabe and Richard T. Wetherald analyzed the effects of doubling CO_2 on atmospheric circulation (Manabe & Wetherald, 1975). This new trend in research revived the scientific contributions of John Tyndall, who in 1859 had discovered the greenhouse effect produced by carbon dioxide and water vapor (Tyndall, 1861), and Svante A. Arrhenius, who had shown that doubling the concentration of CO_2 could raise the Earth's temperature by several degrees (Arrhenius, 1896).

From the 1980s onward, progress in the development of global climate models has been remarkable, thanks to the exponential growth in computing power and the availability of increasingly advanced instruments for measuring and monitoring climate variables, which allow the validation of models through comparison with observations. During those years, more efficient and accurate numerical integration techniques were also developed, and phenomena previously neglected were incorporated into the models, such as the complex dynamics of atmosphere–ocean interactions, cloud formation, their effect on the radiative energy balance, and the impact of aerosols.

In 1988, the Intergovernmental Panel on Climate Change (IPCC) was established, with the aim of studying global warming. The IPCC does not directly conduct research, monitoring, or data collection; rather, it bases its assessments mainly on published scientific literature and on reports from the world's leading institutions.

- Assess the ability of different GCMs to reproduce observed past climate variability and trends, and to analyze their responses to external forcing;
- Characterize the climate system response under plausible future forcing trajectories;
- Investigate the predictability of the climate system across temporal and spatial scales and produce predictions initialized from observed climate states while quantifying and exploring the sources of uncertainty in climate projections.

Within CMIP, simulations from multiple GCMs, developed by numerous research centers worldwide, are evaluated using a standardized set of greenhouse gas emission scenarios (O'Neill et al., 2016). These scenarios are designed to reflect key uncertainties associated with alternative future socioeconomic and technological pathways. CMIP outputs are therefore essential for studies assessing the environmental, social, and economic impacts

Table 4.3 Set of climate scenarios from the CMIP6 project

Shared socioeconomic pathways (SSPs)	*2100 Forcing level W m^{-2}*	
	TIER 1 first priority level	*TIER 2 second priority level*
SSP1 - Sustainability	2.6 (+LTE)	1.9
SSP2 - Middle of the road	4.5	
SSP3 – Regional rivalry	7 (+ ES)	
SSP4 - Inequality		3.4 (OS+LTE)/6
SSP5 - Fossil-fueled development	3.4 (OS)	8.5 (+LTE)

Source: O'Neill et al. (2017).Note: OS Overshoot, LTE Long-Term Extension, ENS Initial Condition Ensemble.

of climate change, as well as for informing mitigation and adaptation strategies. They also represent a major input to assessments conducted by the IPCC.

At the same time, CMIP scenarios provide a common basis for addressing a broad range of scientific questions, including the role of specific forcings such as land-use change and aerosols, the contribution of internal natural variability, and the socioeconomic implications of pathways consistent with limiting global warming to below 2°C (Riahi et al., 2017). The CMIP6 scenarios, organized into two priority tiers, are shown in Table 4.3.

Overall, the general structure of the CMIP6 scenario grid includes, at the first level, four socioeconomic pathways (Shared Socioeconomic Pathways, SSPs), which update and extend those adopted in CMIP5 and are combined with prescribed levels of anthropogenic radiative forcing expressed through the Representative Concentration Pathways (RCPs) having anthropogenic radiative forcing at 2100 equal to 2.6,4.5,6 and 8.5 W m^{-2}, respectively. At the second level, 4 additional SSP combinations are used to explore policy-relevant pathways, and overshoot dynamics are included, designed to explore specific dimensions of future climate change.

A major innovation introduced in CMIP6 is the SSP1–1.9 scenario, which is consistent with the objectives of the Paris Agreement and, in particular, with limiting global warming to 1.5°C above the pre-industrial level.

Each scenario shown in Table 4.3 is associated with corresponding projections of greenhouse gas emissions and land-use change, as illustrated in Figure 4.14. These trajectories are derived from IAMs and are used within IPCC assessments to define mitigation pathways consistent with achieving the Paris Agreement targets (Rogelj et al., 2018).

By incorporating mitigation pathways derived from IAMs into GCM-based simulations, CMIP6 enables a direct link between the large volume of

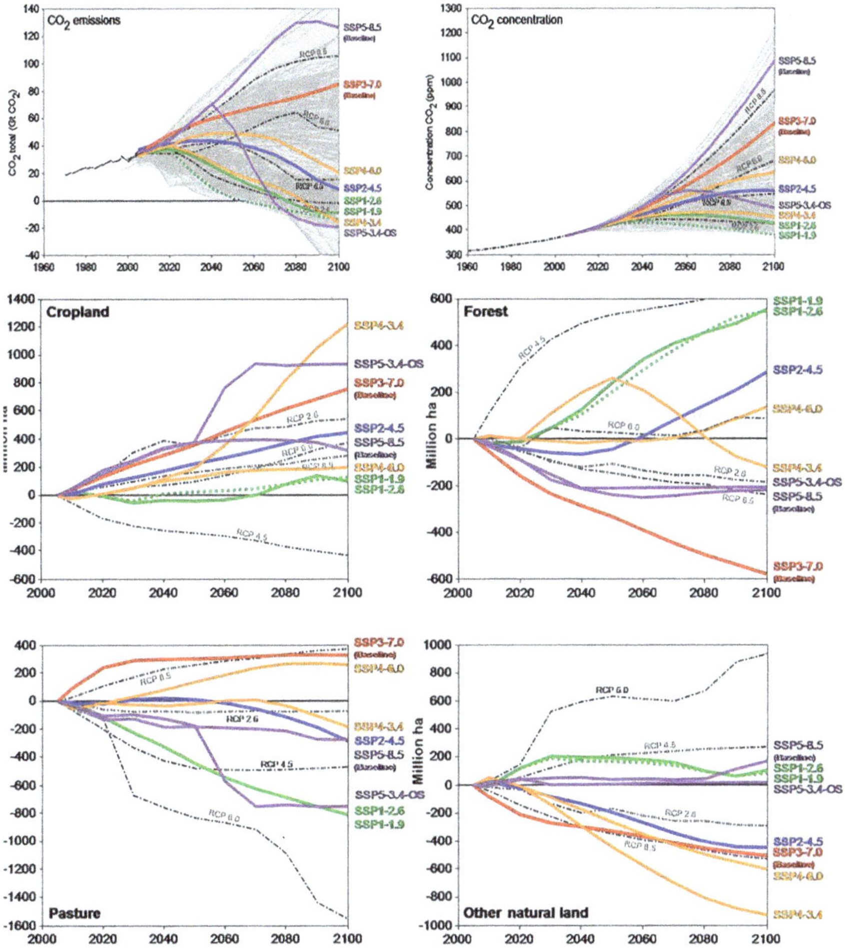

Figure 4.14 Greenhouse gas emissions and land-use changes associated with CMIP6 scenarios.

Source: O'Neill et al., 2017.

output produced by climate models, namely, the three-dimensional spatio-temporal evolution of key climate variables such as wind fields, ocean currents, temperature, humidity, pressure, and rainfall and snowfall, and the socioeconomic, technological, and land-use policies designed to limit global warming. In doing so, the project provides climate information consistent with different shared mitigation pathways, supporting research communities engaged in quantifying local-scale vulnerability across specific regions and, ultimately, in evaluating the effectiveness of adaptation strategies.

4.5.4 Climate projections

About 40 different research groups participate in the CMIP project, each with its own GCM that differs in representation, formalization, spatial resolution (generally varying between 1° and 2.5° latitude), integration technique, and parameterization of sub-grid scale models. The GCM models, validated through comparison with observations, have, over the years, enabled simulations of past and present climate and future projections for the SSP and RPC scenarios, discussed in the previous section.

This represents an enormous amount of data, organized in standard formats, which the entire scientific community can freely access to investigate the models' ability to represent specific climate-related phenomena, such as the reproduction of certain large-scale atmospheric modes like ENSO or the North Atlantic Oscillation (NAO), or other phenomena like the temporal and spatial variability of monsoons, changes in rainfall regimes in different parts of the world, and changes in large-scale ocean circulation (AMOC).

Focusing on the variables that directly concern the hydrological cycle, one can refer to the results of simulations performed with a few dozen GCM models participating in the CMIP6 project (Tebaldi et al., 2021), regarding the trends in global atmospheric surface temperatures and precipitation shown in Figure 4.15.

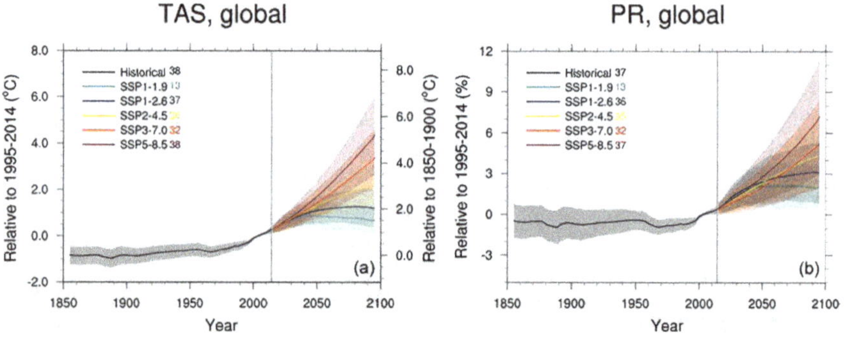

Figure 4.15 (a) Time series of global mean temperature (11-year running means) changes relative to the current baseline (1995–2014, left axis) and the pre-industrial baseline (1850–1900, right axis, obtained by adding an offset of 0.84∘C for SSP1–1.9, SSP1–2.6, SSP2–4.5, SSP3–7.0, and SSP5–8.5); (b) Time series of global mean precipitation (11-year running means) as percentage changes relative to the current baseline (1995–2014) for SSP1–1.9, SSP1–2.6, SSP2–4.5, SSP3–7.0, and SSP5–8.5. The solid lines represent the ensemble means of the temperature trends from simulations performed for the number of models indicated in the legend. The shading represents the 5–95% confidence interval around the mean of a normal distribution.

Source: Tebaldi et al., 2021.

By the end of the century (i.e., as an average over the period 2081–2100), the range of atmospheric surface temperature increase simulated across the set of scenarios, also accounting for the uncertainties associated with each simulation (5 and 95% confidence bands), is projected to be between 0.89 and 6.4°C relative to the 1850–1900 period. The smallest temperature increase corresponds to the SSP1–1.9 mitigation scenario, which represents the lower bound of the projected outcomes and entails a rapid decline in greenhouse gas emissions to zero, followed by net negative emissions sustained over a long period. Only the SSP1–2.6 and SSP1–1.9 scenarios are consistent with meeting the Paris Agreement objective of limiting global temperature increase to below 2°C.

Global precipitation is projected to increase, in agreement with the Clausius–Clapeyron relationship. As shown in the figure, the average change in global precipitation across all scenarios by 2081–2100 (expressed as a percentage relative to the 1995–2014 baseline) ranges between 2.0 and 3.0% for SSP1–1.9 and SSP1–2.6 (the scenarios with the lowest radiative forcing), between 4.2 and 4.9% for SSP2–4.5 and SSP3–7.0, and reaches 7.3% under SSP5–8.5.

Uncertainty in precipitation projections is greater than that associated with temperature projections, owing to differences among models in the mathematical representation of hydrological processes and in the parameterization of sub-grid-scale phenomena. Comparisons with observational data indicate that CMIP5 and CMIP6 simulations for the historical period exhibit systematic biases of varying magnitude depending on the region considered (Kim et al., 2020), particularly in tropical areas. Nevertheless, changes in the probability distribution of precipitation can still be assessed with a sufficient degree of reliability, as can future trends in rainfall regimes across different regions of the globe.

As shown in Figure 4.16, which reports the spatial distributions of temperature and precipitation increases in the last two decades of the 21st century compared to the 1995–2014 period, the variations are not spatially uniform.

Across all scenarios, a more pronounced temperature increase is projected at high latitudes of the Northern Hemisphere, reaching approximately 3°C under the SSP1–1.9 scenario and exceeding 9°C under SSP5–8.5, particularly over continental regions. This leads to a substantial modification of the pole-to-equator and land–ocean temperature gradients, which represent the primary drivers of the climate system, and consequently to changes in the structure of atmospheric circulation. In particular, a northward shift of the jet stream, an intensification of the Hadley circulation, and significant alterations in monsoon systems are consistently projected (Wang et al., 2020). These changes are expected to influence the global hydrological cycle, as illustrated in the figure, resulting in marked decreases in precipitation over regions such as the Mediterranean, Amazonia, Central and Western America,

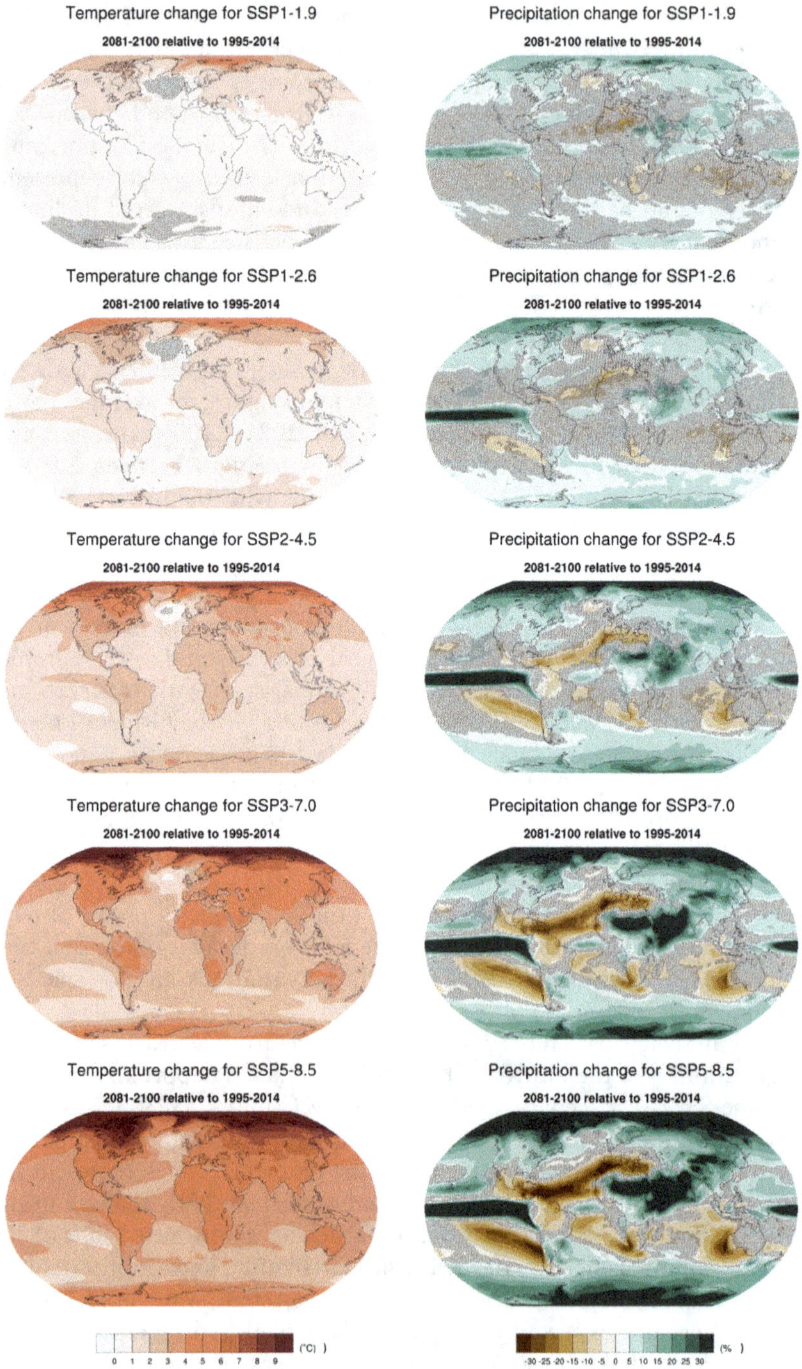

Figure 4.16 Spatial distribution of temperature (∘C) and precipitation (%) changes for the five CMIP6 scenarios.

Source: Tebaldi et al., 2021.

Southern Africa, and Western Australia and, conversely, increases over Northern Europe and Asia.

The changes in the hydrological cycle, evident in the spatial distribution of mean precipitation shown in Figure 4.16, also have implications for the intensity and frequency of extreme hydrological events, including floods and droughts. Analyses of simulations from multiple CMIP6 models indicate that, during the period 2080–2100, the risk of extreme droughts, defined as prolonged periods with precipitation below the 10th percentile of the 1850–1900 reference period, is projected to increase globally, except for a limited number of regions. In Southern Europe, the Mediterranean, Central America, Western South America, South Africa, and Southern Australia, this risk is projected to become three to four times more likely.

Although drought severity depends on the scenario considered, simulations indicate an increased probability of occurrence in the regions listed above even under scenarios that assume strong mitigation efforts (Cook et al., 2020). This suggests that mitigation alone is insufficient to offset the adverse impacts of climate change and must be complemented by adaptation strategies aimed at enhancing regional resilience to extreme hydrological events.

Regarding extreme precipitation events capable of producing floods, CMIP6 projections show a general increase in the intensity of daily rainfall relative to mean conditions, reflecting a shift in the probability density function of daily precipitation toward more intense events (Scoccimarro & Gualdi, 2020). The higher likelihood of extreme precipitation, even in regions experiencing a decrease in mean rainfall, is consistent with the increased moisture-holding capacity of a warmer atmosphere. There is also broad agreement on a future increase in the frequency and intensity of tropical cyclones, leading to enhanced precipitation and stronger, more destructive winds (Emanuel, 2021).

The growing likelihood of extreme hydrological events, together with processes closely linked to the hydrological cycle, such as sea-level rise driven by ocean thermal expansion and glacier melt, further exacerbates the vulnerability of coastal regions worldwide. Recent analyses (Hermans et al., 2021) project a global mean sea-level rise by 2100 of 0.52–1.05 m for SSP5–8.5, 0.38–0.76 m for SSP2–4.5, and 0.30–0.64 m for SSP1–2.6, based on the 5–95% confidence interval.

4.5.5 The challenge of climate change to water sustainability

As stated in Section 4.4, in the absence of climate change induced by anthropogenic greenhouse gas emissions, the problem of water sustainability essentially consists of identifying water-use policies that ensure the balance of the annual or multi-annual hydrological budget at the local basin or regional scale, or more generally at the scale of multiple hydraulically or hydrogeologically interconnected basins. The identification of such a balance is made possible by the assumption, justified by the substantial climatic stability during the Holocene, that meteorological inputs to a given hydrological unit, although variable over time, are statistically stationary. In other words, they can be described by probability density functions whose statistical moments (mean, variance, etc.) are assumed to be invariant over time. This assumption of statistical stationarity is also commonly applied in the analysis of extreme hydrological events through the definition of the return period, defined as the inverse of the probability of occurrence of a given event. The return period is typically used as a criterion for determining the design values of hydraulic infrastructure.

Therefore, in the absence of phenomena associated with climate change, the problem of water sustainability can be considered "local" (at the scale of one or more interconnected basins) and "statistically stationary," since a one-way coupling from the atmospheric to terrestrial branch of the hydrological cycle is assumed to hold.

As highlighted in the previous sections, anthropogenic greenhouse gas emissions influence, through the non-uniform increase in temperatures over continental areas and oceans, ice melt, and the increase in atmospheric water vapor, the evolution of climate at the planetary scale and, consequently, alter the structure of the hydrological cycle, which is a fundamental component of the climate system itself. Such alterations manifest as increased irregularity of precipitation regimes, enhanced evaporation, reduced soil moisture, and more intense and more frequent extreme hydrological events such as floods and droughts. Although these phenomena occur on the local scale, their driving forces operate at the planetary scale.

Analyses of past climate records, current observations, and global climate model simulations indicate that the climate is evolving along trajectories that deviate significantly from those of the Holocene, the geological epoch during which human civilization developed and which was characterized by relatively stable climatic conditions. This deviation is clearly illustrated in Figure 4.17 (Friedlingstein et al., 2025a).

Due to the imbalance between anthropogenic CO_2 sources and natural sink rates discussed in Section 4.3.1, and in the absence of significant reductions in atmospheric greenhouse gas emissions, climate change processes are intrinsically nonstationary. This is evidenced by the increase in CO_2 atmospheric concentration and the increasing and spatially non-uniform rate of global temperature change over time, which in turn affects precipitation regimes, river streamflow, and groundwater recharge across different regions of the Earth.

Accounting for such non-stationarity to assess future water availability or to define hydrological design variables for infrastructure sizing within the framework of water sustainability is one of the major challenges posed by climate change. This challenge is further compounded by uncertainties related to unknown future greenhouse gas emission scenarios.

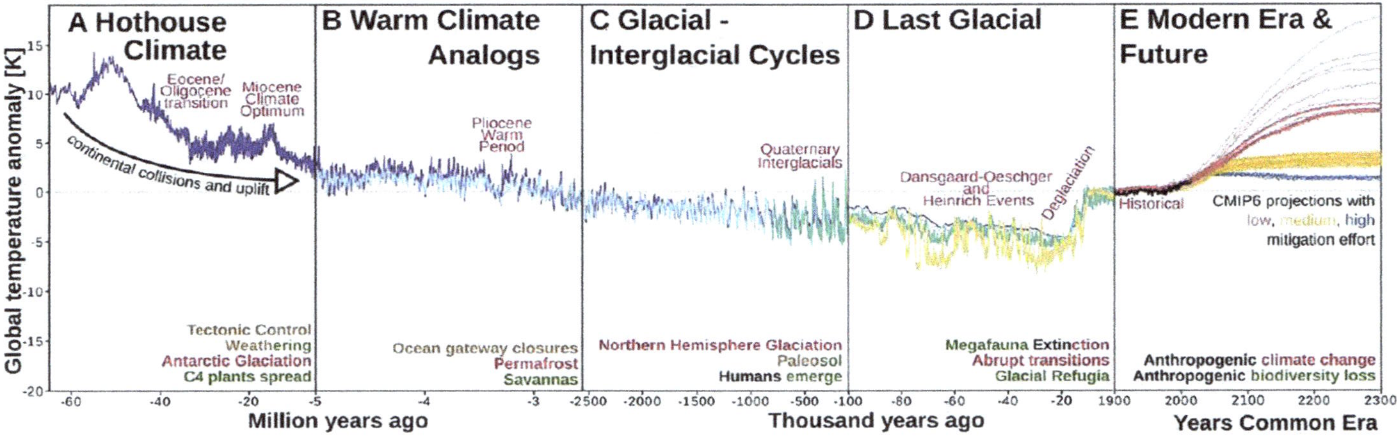

Figure 4.17 Global mean temperature evolution over the last 60 million years BP to the year 2300 based on paleoclimate reconstructions (panels A, B, C, D), in the context of observations and modeled projections for high (SSP1–1.9), medium (SSP2–45), and low (SSP5–85) mitigation efforts (panel E).

Data and calibration details can be found in Rehfeld and Ziegler (2023).

(Source: Friedlingstein et al., 2025a).

Thus, climate change introduces uncertainty and non-stationarity about the future alterations of the hydrological cycle, undermining traditional assumptions of stable flood frequencies, drought statistics, and water availability. Flood design must therefore shift from fixed return periods to risk-based and adaptable approaches, while drought planning should rely on scenario-based analyses and flexible management triggers. Water allocation systems need to move from fixed entitlements to dynamic, availability-based rules that explicitly share risk among users. Overall, effective water management under climate change requires robust, adaptive strategies capable of performing across a wide range of plausible future conditions. The shift toward risk-based and adaptive water management thus represents a core climate change adaptation strategy in response to hydrological non-stationarity. By promoting flexibility, efficiency, and nature-based solutions, such approaches can also indirectly support mitigation by avoiding maladaptive, energy-intensive responses (Merz et al., 2014).

Furthermore, the sustainability of the water cycle must consider all the temporal and spatial scales that characterize it in relation to the various issues of water management (Nese & Grenci 2011).

As emphasized at the end of Section 4.3, interactions and feedback mechanisms between surface and atmospheric processes lead to nonlinear responses that significantly alter the water cycle. For example, urbanization, large reservoirs, and deforestation modify soil moisture and evapotranspiration, which in turn influence rainfall patterns, including in remote regions (Van der Ent et al., 2010). Furthermore, as discussed earlier, water plays multiple planetary functions, such as hydro-climatic and hydro-ecological roles, that are increasingly affected by human activities.

Framing hydrological cycle sustainability by explicitly accounting for the hydro-ecological, hydro-climatic, storage, and transport functions of water at the planetary scale further reinforces the need for a conceptual revision of water sustainability (Richardson et al., 2023; Wang-Erlandsson et al., 2022). In this new context, where human activities directly influence the evolution of the Earth system, the concept of water sustainability acquires broader and more complex meanings. It must be embedded within the wider framework of climate change and its social, economic, and environmental impacts (Gleick, 2003b). This paradigm shift has direct implications for water governance and for achieving the SDGs mentioned in Section 4.2, calling for integrated water policies consistent with the biophysical constraints of the Earth system.

Which practices can be adopted to limit undesirable alterations of the global water cycle that may prove environmentally, socially, and economically unsustainable?

This question leads to the broader issue of identifying limits or threshold values of climatic variables beyond which impacts on global environmental, social, and economic systems become unsustainable. In the case of global temperature increase, for example, such limits are related to greenhouse gas emission rates, which form the basis for defining the different scenarios of the CMIP6 project (O'Neill et al., 2017). Other variables associated with tipping elements, discussed in Section 4.5.2 and shown in Table 4.1, can also be used to define threshold values (tipping points), below which, based on current scientific understanding, positive feedback mechanisms that could amplify undesirable effects, such as disproportionate temperature increases, are not triggered (Lenton et al., 2008; Lenton & Williams, 2013; Schellnhuber, 2009).

Anthropogenic impacts on the hydrological cycle cannot be attributed solely to greenhouse gas emissions. Land-use change, urbanization, and the overexploitation of surface and groundwater resources for domestic, industrial, agricultural, and energy purposes alter both the quantity and quality of freshwater resources upon which humanity depends, physically and economically. These pressures also directly affect natural habitats and ecosystems that rely on the availability of water in adequate quantity and quality (Cosgrove & Loucks, 2015).

In an effort to define a safe operating space for humanity (Rockstrom et al., 2009a), a group of 29 researchers published a seminal article in *Nature* in 2009 identifying nine planetary boundaries that should not be exceeded to avoid unacceptable risks. These boundaries include climate change, ocean acidification, biodiversity loss, biogeochemical flows of nitrogen and phosphorus, stratospheric ozone depletion, global freshwater use, land-use change, atmospheric aerosol loading, and chemical pollution. For each boundary, one or more control variables are defined, with associated numerical thresholds delimiting the safe operating space. The 2°C temperature limit, later formally recognized in the Paris Agreement, is conceptually similar to that proposed by Rockström and colleagues. The updated planetary boundaries framework published in Rockström et al. (2024) identifies freshwater change as one of the boundaries that has already been exceeded at the global scale.

Recognizing that a planetary boundary based solely on global freshwater consumption cannot capture the complexity of water's role in the Earth system, Gleeson and colleagues proposed six planetary sub-boundaries addressing the different water compartments and functions within the Earth system (Gleeson et al., 2020b), as discussed in Section 4.4. These sub-boundaries include:

1. Evapotranspiration, crucial for land–atmosphere coupling (hydro-climatic regulation), is related to atmospheric water.
2. Precipitation patterns sustaining biomes and biodiversity (hydro-ecological regulation) are also linked to atmospheric water.
3. Carbon uptake or net primary productivity is dependent on soil moisture.
4. Hydraulic characteristics of rivers and lakes (e.g., discharge, residence time) are essential for aquatic biodiversity and surface waters.

5. River baseflows supplied by groundwater discharge are fundamental for ecosystem maintenance.
6. Ice sheet volume, which plays a critical role in sea-level rise, is associated with water stored in glaciers and polar regions.

Table 4.4 summarizes, for each sub-boundary, the relevant spatial scales, potential control variables, and response variables for monitoring changes. In particular, Gleeson et al. (2020b) propose defining these sub-boundaries based on future projections consistent with greenhouse gas emission pathways that comply with the Paris Agreement, namely, limiting global temperature increase to below 2°C relative to pre-industrial levels.

The planetary boundaries defined in this way collectively constitute a global-scale framework, consistent with the challenges posed by the climate emergency, within which existing and more traditional water resource management and water cycle sustainability practices at the local (basin), regional, or national scales can be integrated (Zipper et al., 2020). This change in perspective involves not only scientific aspects but also governance at both national and international levels, which must address their efforts in favor of mitigation and adaptation actions needed to cope with climate change and its impact on the hydrological cycle.

4.6 ACTIONS TO COPE WITH CLIMATE CHANGE: MITIGATION AND ADAPTATION

The United Nations Framework Convention on Climate Change (UNFCCC) identifies two strategies to address climate change: mitigation and adaptation. These represent the core elements of the Paris Agreement of December 12, 2015, whose main goal is to strengthen the collective global response of nations to keep the rise in global temperature in the 21st century well below 2°C above pre-industrial levels, while pursuing efforts to limit it to 1.5°C. These threshold values are justified by the considerations already outlined, such as an increase in temperature beyond 2°C:

- would result in average values of temperature higher than those of the past 800,000 years, especially compared to those of the Holocene;
- could trigger tipping elements of the climate system that would lead the planet's climate along evolutionary pathways characterized by environmentally, socially, and economically unsustainable conditions (Steffen et al., 2018).

Mitigation refers to the causes of global warming and is aimed at limiting the temperature increase by implementing appropriate policies to reduce greenhouse gas emissions into the atmosphere. Adaptation refers to the ways people respond to the consequences of global warming. It includes all

Table 4.4 Suggestions for key aspects of each of the six sub-boundaries, including possible interim planetary boundary based on Paris Agreement 2_C target for late this century

	Atmospheric Water (Hydroclimate)	*Atmospheric Water (Hydroecology)*	*Soil Moisture (Hydroclimate)*	*Surface Water (Hydroecology)*	*Groundwater (Hydroecology)*	*Frozen Water (Storage)*
Possible scale of analysis	Distributed hydrological unit	Biomes or hydro-climatic regimes	Biomes or land cover groups	Large basins or river networks	Regional aquifers	Global
Possible response variables	Climate pattern stability or land-atmosphere coupling stability	Terrestrial biosphere integrity	Carbon uptake or net primary productivity	Aquatic biosphere integrity (species richness or species/area)	Terrestrial or aquatic biosphere integrity	Sea-level rise
Possible interim planetary boundary	Percentage of global land area with evapotranspiration change within the range of simulated future	Percentage of global land area with precipitation change within the range of simulated future	Maintenance of global net primary productivity at or above levels under simulated future	Percentage of basins or total river length within environmental flow limits under simulated future	Percentage of basins with low flows meeting or exceeding the simulated future	Volume of ice melt to keep sea level within limits under simulated future

The key Earth System functions of water for each sub-boundary are identified in parentheses (such as hydroecology for surface water).

Source: Gleeson et al. (2020b).

technical, social, and economic actions aimed at reducing the vulnerability of specific areas of the planet to the impacts of climate change and increasing their resilience, that is, the capacity of a given social, economic, or environmental system to absorb disturbances without altering its stability characteristics. An important difference between mitigation and adaptation is that the former is conceived at the global scale, while the latter refers to the local scale of specific geographic areas.

Both strategies must be considered as unified and complementary. In principle, the more effective mitigation policies are, the fewer adaptation actions are required for a given area. However, uncertainties regarding the effectiveness of mitigation actions, partly due to difficulties in international governance of national political and economic actions and partly to uncertainties in our projections of the future climate arising from the limitations of current forecasting tools, suggest the need to plan adaptation policies now to increase the resilience of specific areas. This observation is supported by the fact that the effects of climate change are already evident today, such as the intensification and greater frequency of extreme hydrological and climatic events, including droughts, floods, heat waves, and wildfires.

Given the inertia of the Earth system, climate change is expected to continue for decades, if not centuries, even under efficient mitigation policies consistent with the Paris Agreement targets. Global temperature has already risen by 1.1°C since 1880, and the rate of warming is the fastest in the past 65 million years. Since climate risks manifest at the local scale, and not all regions of the planet have the same vulnerability to climate change impacts, their effects must be assessed within a specific geographical context, and appropriate adaptation policies must be adopted.

There is a clear thematic coherence between many of the objectives of the Paris Agreement and the content of the 2030 Agenda for Sustainable Development (Janetschek et al., 2020). Consequently, the Agenda's sustainability goals effectively constitute a roadmap which, if implemented, would in themselves create conditions of greater resilience to the adverse local effects of climate change. In confirmation of this, Table 4.5 presents the main characteristics of the five shared socioeconomic pathways (already shown in Table 4.3; O'Neill et al., 2017), which clearly demonstrate that sustainable development reduces the challenges associated with both mitigation and adaptation to climate change.

It is likely that socioeconomic impacts at the local scale will propagate non-linearly once climate risk exceeds a certain threshold, beyond which the social, economic, and ecological systems will evolve toward a state of irreversible crisis. Therefore, implementing sustainability goals would move the system away from this threshold. Although the direct impact of climate change is local, the strong interconnectedness of different socioeconomic and financial systems can generate significant effects even in regions far removed from the area experiencing the direct impact. There is growing evidence that changing environmental and climate conditions are

Table 4.5 Key characteristics of the five shared socioeconomic pathways (SSPs)

Socio-Economic Challenges to Mitigation	*Socio-economic Challenges to Adaptation*		
	Low	*Medium*	*High*
High	SSP5: Fossil-fueled development • Low population • Very high economic growth per capita • High human development • High technological progress • Ample fossil fuel resources • Very resource intensive lifestyles • High energy and food demand per capita • Economic convergence and global cooperation		SSP3: Regional rivalry • High population • Low economic growth per capita • Low human development • Low technological progress • Resource-intensive lifestyles • Resource-constrained energy and food demand per capita • Focus on regional food and energy security • Regionalization and lack of global cooperation
Medium		SSP2: Middle of the road • Medium population • Medium and uneven economic growth • Medium and uneven human development • Medium and uneven technological progress • Resource-intensive lifestyles • Medium and uneven energy and food demand per capita • Limited global cooperation and economic convergence	

(*Continued*)

Table 4.5 (Continued)

Socio-Economic Challenges to Mitigation	*Socio-economic Challenges to Adaptation*		
	Low	*Medium*	*High*
Low	SSP1: Sustainable development • Low population • High economic growth per capita • High human development • High technological progress • Environmentally oriented technological and behavioral change • Resource-efficient lifestyles • Low energy and food demand per capita • Economic convergence and global cooperation		SSP4: Inequality • Medium to high population • Unequal low to medium economic growth per capita • Unequal low to medium human development • Unequal technological progress: high in globalized high-tech sectors, slow in domestic sectors • Unequal lifestyles and energy /food consumption: resource intensity depending on income • Globally connected elite, disconnected domestic work forces

Source: O'Neil et al. (2017).

exacerbating human migration, which already affects vulnerable sections of humanity (Balsari et al., 2020). This should encourage the pursuit of sustainability goals and the implementation of adaptation policies that involve social, economic, and ecological systems worldwide according to a universal and inclusive logic.

4.6.1 Climate change mitigation

Climate change mitigation consists of reducing or preventing the emission of greenhouse gases. This may require a broader use of renewable energy, the adoption of new technologies, the upgrading of older equipment to make it more energy efficient, changes in agricultural management practices, or shifts in consumer behavior. Mitigation can take the form of complex operations, such as a mitigation plan for a city, region, or nation, or simple improvements, such as enhancements to the design of a cooking stove.

Climate change mitigation is closely tied to the concept of the carbon budget, or, more precisely, the remaining carbon budget, the finite total amount of CO_2 that can be emitted into the atmosphere by human activities while still keeping global warming within a desired temperature limit.

The concept of the carbon budget was introduced by Matthews et al. (2009), who demonstrated the existence of an almost perfectly linear relationship between the increase in the planet's global surface temperature, relative to the 1850–1900 average, and the cumulative emissions of fossil fuels since 1876, expressed in $GtCO_2$. The slope of the regression line is called the Transient Climate Response to Cumulative Emissions (TCRE), a term widely used in scientific literature.

Carbon budgets have become a useful tool for communicating the challenges we face in keeping warming to 1.5°C and well below 2°C, the limits on the rise of average global temperature established by the Paris Agreement.

As reported in the IPCC Special Report (Allen et al. 2018) on the social, economic, and environmental impacts of limiting surface temperature increases to 1.5°C above pre-industrial levels (Rogelj et al., 2018), it can be inferred that, with an estimated temperature increase of 0.87 ± 0.12°C from 1850–1900 to 2006–2015, only an additional 0.63°C increase is available if the global temperature rise is to be kept below 1.5°C compared to the pre-industrial era.

On the basis of such calculation, the carbon budget that can still be emitted to limit the increase of global temperature to 1.5 °C as of January 1, 2018, can be estimated at 840, 580, and 420 $GtCO_2$, corresponding to the 33rd, 50th, and 67th percentiles of the TCRE, respectively. Taking into account various sources of uncertainty, variations in the remaining carbon budget of up to ±50% are possible.

It is important to note that a remaining carbon budget of 580 $GtCO_2$ (50th percentile of the TCRE), assuming a linear decline in CO_2 emissions, requires achieving net-zero emissions within the next thirdly years, a time frame that is significantly reduced in the case of smaller remaining carbon budgets. Despite the robustness of the remaining carbon budget concept from a scientific perspective, there remain significant uncertainties in estimating this quantity. In any case, to limit the rise in global temperature, net emissions must reach zero at some point in the near future.

Various pathways can be envisaged to keep temperature increases below 1.5 or 2°C, depending on how humanity will use energy in the future, the technologies that will be developed, demographic factors, and the organization of societies. A comprehensive discussion of these factors is provided in the aforementioned IPCC report, which anticipates large-scale, radical transformations in energy production systems, agriculture, land use, and the ways in which food, energy, and materials are consumed.

Even if humanity succeeds in limiting the temperature increase to 1.5 or 2°C over the next few decades, it will not automatically guarantee that the effects of climate change will not manifest. Table 4.6 highlights the global negative impacts, most of which are related to water cycle issues, associated with two scenarios aimed at keeping the global temperature below the 1.5 and 2°C thresholds, respectively. It should be noted that climate change impacts worsen significantly under the 2°C mitigation scenario. Table 4.6 shows how the 1.5°C scenario allows for a more achievable path and how these impacts make it more difficult to meet the sustainability goals of the 2030 Agenda. At the same time, it is important to emphasize that achieving the Agenda 2030 goals enhances the resilience of social, economic, and environmental systems to climate change. This means that the goals of the 2030 Agenda and climate mitigation, through the reduction of greenhouse gas emissions, must be pursued simultaneously.

4.6.2 Adaptation to climate change

Climate change adaptation consists of identifying, planning, and implementing policies or infrastructure aimed at limiting the adverse effects of expected future climate change on specific geographical areas.

The adverse impacts of alterations in the hydrological cycle due to climate change manifest through water scarcity, degradation of water quality, and an increased frequency and intensity of extreme hydrological events, as well as sea-level rise in coastal areas. This phenomenon further exacerbates the social, economic, and environmental impacts of floods and droughts in these regions.

Effective flood-risk adaptation requires a diversified set of interventions, which may include structural flood-protection measures, rapid early warning systems, adequate land-use planning that takes flood risks into account, the restoration of floodplains to allow the natural expansion of floods, the development of social protection systems, and financial interventions (Jongman, 2018; Merz et al., 2014).

Adaptation to droughts relates to the problem of water scarcity for various uses, particularly agriculture, resulting from reduced volumes of water

Table 4.6 Sustainable development implications of avoided impacts between 1.5 and 2°C global warming

Impacts	*1.5°C*	*2°C*	*Sustainable Development Goals (SDGs) More Easily Achieved when Limiting Warming to 1.5°C*
Water scarcity	4% more people exposed to water stress	8% more people exposed to water stress, with 184–270 million people more exposed	SDG 6 water availability for all
	496 (range 103–1159) million people exposed and vulnerable to water stress	586 (range 115–1347) million people exposed and vulnerable to water stress	
Ecosystems	Around 7% of land area experiences biome shifts	Around 13% (range 8–20%) of land area experiences biome shifts	SDG 15 to protect terrestrial ecosystems and halt biodiversity loss
	70–90% of coral reefs at risk from bleaching	99% of coral reefs at risk from bleaching	
Coastal cities	31–69 million people exposed to coastal flooding	32–79 million exposed to coastal flooding	SDG 11 to make cities and human settlements safe and resilient
	Fewer cities and coasts exposed to sea-level rise and extreme events	More people and cities exposed to flooding	
Food systems	Significant declines in crop yields avoided, some yields may increase	Average crop yields decline	SDG 2 end hunger and achieve food security
	32–36 million people exposed to lower yields	330–396 million people exposed to lower yields	
Health	Lower risk of temperature-related morbidity and smaller mosquito range	Higher risks of temperature-related morbidity and mortality and larger geographic range of mosquitoes	SDG 3 to ensure healthy lives for all
	3546–4508 million people exposed to heat waves	5417–6710 million people exposed to heat waves	

Source: Roy et al. (2018).

available in lakes, reservoirs, or groundwater aquifers. The latter, when located in coastal areas, may be subject to saltwater intrusion, exacerbated by sea-level rise, with a consequent decline in water quality.

Policies for drought adaptation can involve a wide range of practices, including the construction of new infrastructure to increase storage capacity in surface and groundwater bodies or to prevent saltwater intrusion; improving the efficiency of distribution networks; introducing agricultural practices that reduce irrigation water demand; implementing water recycling and reuse practices; and developing seasonal drought-risk forecasting models to manage water resources more effectively. Some authors suggest considering flood- and drought-risk adaptation policies jointly in order to avoid interventions aimed at preventing one risk from producing negative effects on the other (Ward et al., 2020).

In any case, the objective of adaptation actions is to increase the resilience of a specific geographical area by reducing hydraulic risk associated with extreme hydrological events, which can be calculated as the product of three factors:

- Hazard, defined as the probability of occurrence of an event of a given intensity;
- Vulnerability of an element (person, buildings, infrastructures, and economic activities), meaning the propensity to suffer damage as a consequence of stresses induced by an event of a certain intensity;
- Exposure, i.e., the number of elements at risk present in a given area.

(Disse et al., 2020)

Therefore, reducing hydraulic risk through adaptation policies or measures requires the assessment of these three components of risk.

The assessment of the hazard of extreme hydrological events, through the identification of their intensity and frequency, is one of the most problematic issues related to climate change. Global warming modifies the structure of large-scale atmospheric circulation over time, as well as the transport of atmospheric moisture, upon which the local rainfall regime depends. This, in fact, makes it difficult to identify the intensity and frequency of project hydrological variables (such as, for example, the design flood discharge in the case of flood defense infrastructure) based on historical data from previous years, since the natural variability of precipitation patterns is accompanied by the effects of climate change, which results in conditions that vary over time depending on the mitigation policies for anthropogenic forcing that will be concretely implemented.

To capture these variations in the rainfall regime at the scale of the specific geographical area of interest, it is necessary to refer to climate projections by GCMs discussed in Section 4.5.3, which are the only tools capable of simulating future atmospheric/oceanic circulation structures, as well as temperature and precipitation in their spatial and temporal variability.

The direct use of GCM-simulated temperature and precipitation to identify the best adaptation policies for a specific geographical area is problematic, as there are several sources of uncertainty. Firstly, although various possible mitigation pathways, corresponding to different climate scenarios, have been hypothesized in CMIP5 or CMIP6, no one knows which scenario is most probable or what probability to assign to each; a second source of uncertainty is inherent to the different characteristics of the GCMs and their relative simulations: sensitivity to initial conditions, parametric uncertainties, and structural uncertainties influence GCM simulations, which represent paths of climate evolution that can differ significantly even for the same emission scenario (Parker, 2010).

Thirdly, to explore possible adaptation policies at a local or regional scale, it is necessary to project the trends of hydrological variables such as temperature, precipitation, or sea-level rise with a finer spatial resolution than that provided by GCMs, which is coarser and where these variables are often affected by systematic errors. For example, in CMIP5/6GCMs, processes are simulated with a horizontal resolution generally greater than 1 degree of latitude and longitude (about 100 km × 100 km), which is not sufficient to represent the complex morphology of continental areas affecting significantly the rainfall regime at the local scale, as well as the bathymetric variation affecting local phenomena, such as coastal currents (Holt et al., 2009), or small-scale phenomena, such as eddies (Penduff et al., 2011). Structural errors in the models are also likely due to a still incomplete understanding of the phenomena or feedbacks that influence atmospheric and ocean circulation (Little & Urban, 2016), so corrections must be applied to compensate for model distortion. Precipitation and temperatures simulated by GCMs and Regional Circulation Models (RCMs) are generally affected by systematic errors compared to the historical observed values, and, in order for them to be used in quantifying climate projections, bias correction methods are generally applied (Teutschbein & Seibert, 2012).

Dynamic downscaling and statistical methods are used to obtain projections of precipitation, temperature, or sea-level rise at a spatial resolution compatible with the dimensions of specific geographical areas and the necessities imposed by various water management issues. The former consists of using RCMs, which integrate equations analogues the GCM ones but over smaller domains, typically on a regional scale (e.g., the Mediterranean region), with a higher spatial resolution (a few tens of kilometers) (Tapiador et al., 2020), using GCM simulations as boundary conditions.

Alternatively, statistical downscaling methods are employed. These are based on identifying a statistical correlation between the rainfall regime or temperature in a specific area and suitable atmospheric predictors, which are representative of the circulation structure at the synoptic scale (Cioffi et al., 2016, 2017, 2020; Conticello et al., 2018). More recently, machine learning methods have been proposed to transform coarse-scale data into the fine-scale information necessary for hydrological assessment. These methods seem to be very promising in capturing the complex spatial-temporal structure of fine-scale storms, even if they partially fail to accurately reproduce marginal statistics of extremes (Papalexiou & Mamalakis, 2026; Shafei & Cioffi, 2024a,b).

The effects of the various sources of uncertainty mentioned above are shown in Figure 4.18, where it is highlighted that a range of variation, rather than a single value, is associated with the projection of a generic climate variable. In addition to the irreducible uncertainty in projections of future anthropogenic climate change, natural climate variability is also present (Figure 4.18). This variability modulates the climate signal on decadal or multi-decadal scales and partially overlaps with it (Doss-Gollin et al., 2019; Tebaldi & Friedlingstein, 2013). Therefore, integrating climate projections, including their uncertainty, over the operational lifetime of structures or adaptation policies into risk-mitigation decision frameworks is fundamental.

Vulnerability and exposure are the other two components that must be taken into account to reduce the risk due to climate change. Adaptation measures are aimed fundamentally to increase the hydrological resilience to climate change of a given territory or activity through: (a) structural solutions (dams, reservoirs, drainage channels, and flood defense infrastructures);

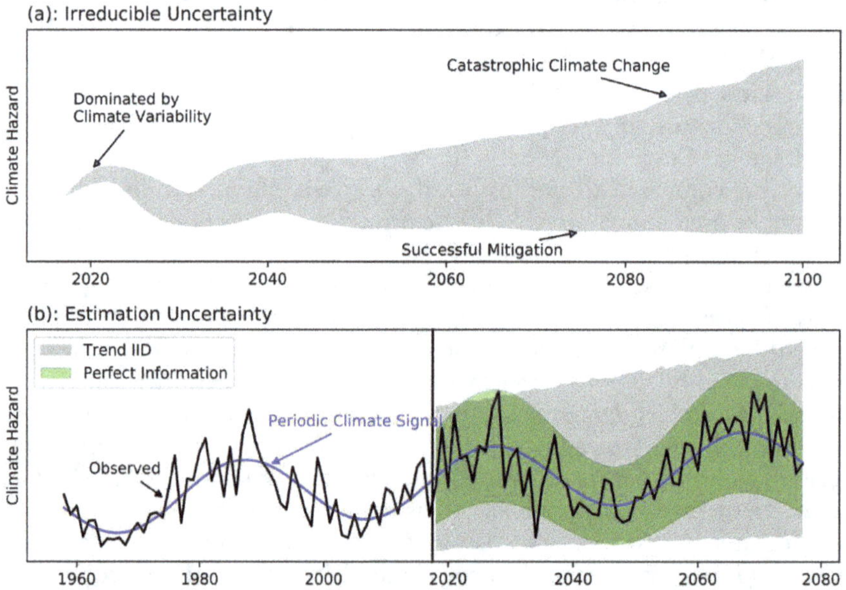

Figure 4.18 Uncertainty in climate projections and natural variability.

(Source: Doss-Gollin et al., 2019.)

(b) ecosystem-based solutions (wetland restoration, reforestation, soil conservation, and reconnection of rivers to floodplains); (c) Long-term climate risk management (monitoring climate change, planning and adapting to future scenarios, integrating water management policies, and involving local communities in risk management) (Granata & Di Nunno, 2026; Disse et al., 2020).

The main steps of a general framework for climate adaptation studies require the identification of: (a) baseline conditions, (b) climate scenarios referred to target years, (c) future territory assets, and (d) best adaptation option among the hypothesized ones in relation to social, environmental, and economic aspects (Elgendy et al., 2024).

Various methodologies have been proposed in the literature in the past for the identification and evaluation of adaptive actions and works to address climate change, their magnitude, and implementation times. These can be grouped into two families: bottom-up (resilience) methodologies and top-down (predictive) methodologies.

Among the bottom-up approaches, an example is the adaptation tipping point (ATP) proposed by Kwadijk et al. (2010). This consists of identifying threshold values for the climate variables of interest (in this specific case, sea-level rise), beyond which a defense work becomes inefficient and must be modified or replaced. According to this approach, various future sea-level rise scenarios are hypothesized, and the works necessary to reduce their impact in terms of the vulnerability and exposure of the affected area are evaluated, along with the corresponding threshold values. Through the monitoring of the climate variable, once a determined threshold value is reached, the defense work is implemented. These approaches do not necessarily require the use of climate projections from GCM models.

The top-down methods consist of designing adaptation strategies as a function of the projections of the intensity and frequency of extreme hydrological events (obtained from downscaled GCM simulations), aimed at increasing the resilience or reducing the vulnerability of a given geographical region to climate change. The approaches mentioned in Chapter 17 of the IPCC WGII (Adger et al., 2007) follow the top-down method; other examples are the methods developed by Muis et al. (2015) and Woodward et al. (2014). Cioffi et al. (2022) have devised an approach that allows for the identification of the most appropriate temporal planning horizons within which to activate adaptation actions or implement flood defense infrastructure (Box 4.6).

4.7 CONCLUSION

The United Nations Sustainable Development Goals Report 2025, published ten years after the adoption of the 2030 Agenda, shows that global progress remains largely off track. Of the 169 SDG targets, only 35% show

BOX 4.6 EXAMPLE OF A TOP-DOWN APPROACH FOR FLOOD DEFENSE

This approach is aimed at identifying the optimal timing for flood defense adaptation actions and the sizing of related works under the climate change scenarios hypothesized in CMIP projects. It specifically considers the uncertainty in future projections of hydrological forcings (rainfall, sea-level rise) and the possibility of abrupt and non-gradual transitions of these forcings, which in principle cannot be excluded (Lenton et al., 2019).

The main element of the top-down approach is a multi-objective optimization model that integrates numerical hydraulic models and damage assessment models to quantify two objective functions that express the optimality criteria:

1. Minimizing flood risk for the entire period of the climate projections.
2. Minimizing the construction and operation cost of the flood defense infrastructure system for the same period.

The formalization of the hydraulic risk associated with flood events is based on the definition of Expected Annual Damage (EAD), integrated over the number of years in the considered projection period. The EAD is calculated as the product of the annual probability of occurrence of extreme hydrological events of a given intensity and the damage produced by the floods associated with them.

The annual probability of occurrence of extreme events is obtained by accounting for uncertainties in future climate projections, which are therefore integrated into the definition of the objective functions. The damage produced by floods is a function of the hydraulic characteristics (depth and velocity) of the flood flow generated by the hydrological forcings, the vulnerability characteristics of buildings and activities present in the area, their exposure, and the state of the flood defense system (e.g., the height of river banks).

As shown in the sketch of Figure 4.19, a hydraulic model is integrated into the multi-objective optimization algorithm for damage assessment. The decision variables for the multi-objective optimization problem are the size of the flood defense structures and the timing of implementation. The input variables are the coupled events of heavy rainfall and storm surges, as well as the mean sea-level rise obtained from climate projections.

The solution to the multi-objective optimization problem consists of a set of optimal solutions, which identify the Pareto Front: this therefore expresses the relationship between the two objective functions (Deb et al., 2002).

Multi-objective optimization methods are more flexible tools in the context of hydraulic risk management decisions because they allow simultaneous consideration of objectives expressed even with non-commensurable metrics, such as those related to the conservation of natural habitats, which cannot be easily expressed in monetary terms (Cioffi and Gallerano, 2012); furthermore, these methods provide decision-makers with a large space of optimal solutions (Pareto set) from which to choose.

adequate progress (18% on track and 17% moderate), while 48% remain insufficient, including 31% with marginal gains and 17% with no progress.

Regarding water-related goals, particularly SDG 6, access to drinking water, sanitation, and hygiene improved between 2015 and 2024. However, in 2024, 2.2 billion people still lacked safely managed drinking water, 3.4 billion lacked safely managed sanitation, and 1.7 billion had no basic hygiene services at home. In schools, 646 million children lacked basic hygiene, requiring a fourfold acceleration to meet 2030 targets. Water systems remain under severe strain from pollution, water stress, weak governance, and limited transboundary cooperation. Only 56% of domestic wastewater is safely treated, and progress in IWRM has been slow, increasing from 49% in 2017 to 57% in 2023. Stronger integration between IWRM and climate adaptation remains largely untapped.

Under SDG 11, rapid urbanization is reshaping development. More than half of the global population now lives in cities, whose water footprint covers 41% of Earth's surface, despite occupying only 2% physically. Urban land in developing countries is projected to triple by 2050. Cities generate around 70% of fossil fuel-related greenhouse gas emissions and must achieve net zero to meet the Paris Agreement. Meanwhile, unplanned urban growth persists: up to 3 billion people struggle to afford adequate housing, and 1.12 billion live in informal settlements without basic services.

Climate change further intensifies risks. Since 1999, global mean sea level has risen by an average of 3.7 mm per year, totaling 9.38 cm over 25 years, with the rate accelerating by 46% (Copernicus Climate Change Service). As many cities are located in coastal areas, rising sea levels are expected to increase coastal erosion, flooding, and saltwater intrusion, amplifying environmental, social, and economic risks.

Finally, SDG 14 (Life Below Water) and SDG 15 (Life on Land) remain under severe pressure due to overfishing, pollution, biodiversity loss, ocean warming and acidification, deforestation, species extinction, and insufficient protection of key ecosystems.

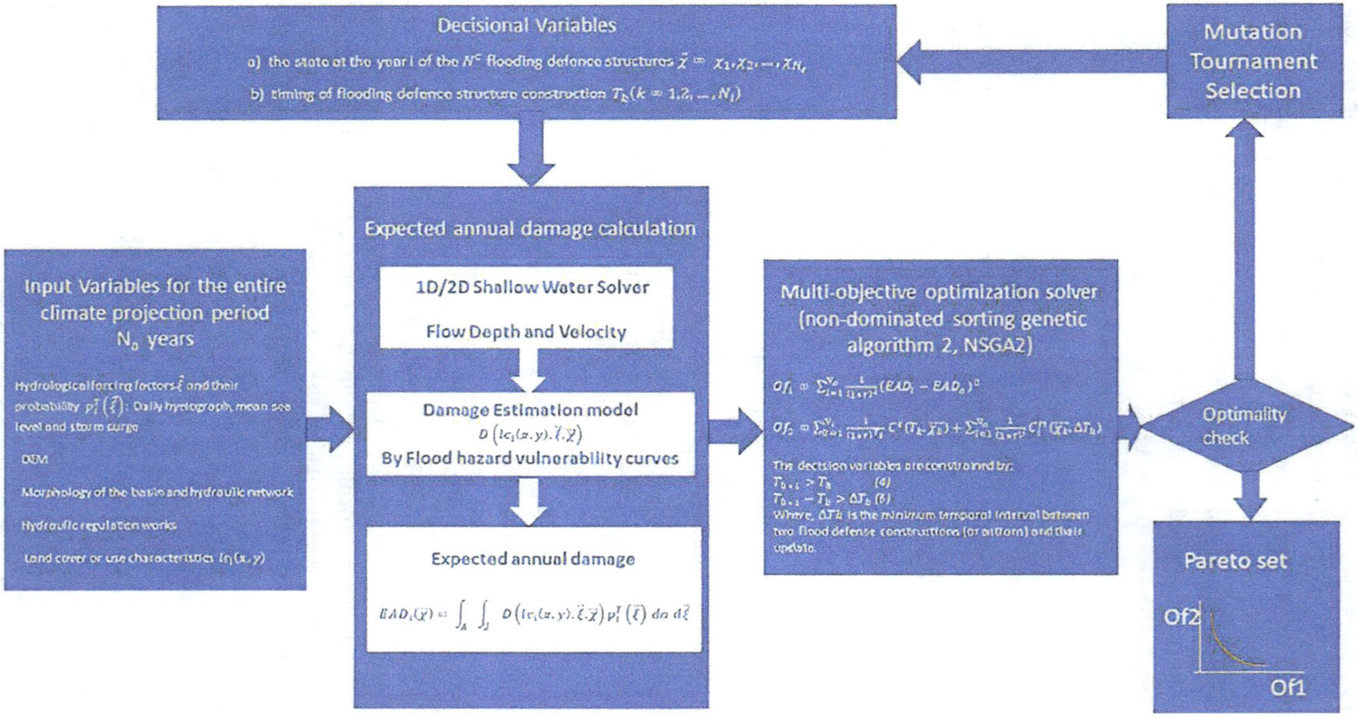

Figure 4.19 Sketch of a multi-objective model for the identification of future flood defense infrastructures.

Climate change is accelerating. In 2024, global temperatures reached approximately 1.55°C above pre-industrial levels, making it the hottest year on record. According to the United Nations Environment Programme's Emissions Gap Report 2024, greenhouse gas emissions reached a record 57.1 gigatons of CO_2 equivalent in 2023, rising by 1.3% from 2022 and exceeding the average annual increase recorded between 2010 and 2019. Under current policies, global warming could reach up to 3.1°C, with severe consequences for societies, ecosystems, and economies. Although disaster-related mortality declined compared to the previous decade, the number of people affected by climate-related disasters and displacement has sharply increased.

Despite progress on the SDGs, growing inequalities and accelerating environmental degradation risk reversing recent gains. Climate change is disrupting the global water cycle, reducing water availability in many regions and intensifying water scarcity, extreme heat, heavy rainfall, floods, droughts, landslides, and sea-level rise. Even limiting warming to 1.5°C could expose 500 million people to severe water stress, 4 billion to heatwaves, and tens of millions to coastal flooding.

As emphasized in the Sixth Assessment Report of the IPCC, the window for effective mitigation is rapidly closing. Global temperatures are already about 1.1°C above pre-industrial levels, and it is highly likely that the 1.5°C threshold identified in the Paris Agreement will be reached within the next two decades under most emissions scenarios. Even with net-zero emissions by 2050, warming could reach around 1.6°C by mid-century, with a high probability of at least one ice-free Arctic summer. Some impacts—including glacier and ice sheet loss, permafrost thaw, ocean warming, acidification, deoxygenation, and long-term sea-level rise—are already irreversible and will persist for centuries or millennia.

In this context, adaptation planning is essential and must align with the economic, social, and environmental objectives of the 2030 Agenda. Adaptation strategies should adopt holistic, multi-scale approaches that address the non-stationarity of hydrological regimes and strengthen both risk management and resilience. Actions must also respect planetary boundaries and local constraints on water use and quality.

While the outlook is critical, solutions exist. Technological and scientific knowledge already enable decarbonization and improved resilience. Early warning systems integrating remote sensing, field data, artificial intelligence, and advanced climate modeling can enhance preparedness and reduce vulnerability, particularly in high-risk regions (Cioffi et al., 2024, 2025).

However, the decisive factor remains strengthened international cooperation, guided by an inclusive vision that reduces inequalities and ensures universal access to safe, high-quality water—an essential foundation for sustainable development.

REFERENCES

Abbott, B. W., Bishop, K., Zarnetske, J. P., Camille Minaudo, F. S., Chapin, S. K., Hannah, D. M., et al. "Human domination of the global water cycle absent from depictions and perceptions." *Nature Geoscience*, 12, no. 7 (2019): 533–540.

Adger, W., Agrawala, S., Mirza, M., Conde, C., O'Brien, K., Pulhin, J., ... and Takahashi, K. *Climate Change 2007: Impacts, Adaptation and Vulnerability*. Cambridge University Press, Contributi, 2007, 717–743.

Allen, M. R., Dube, O. P., Solecki, W., Aragon-Durand, F., Cramer, W., Humphreys, S., et al. "Framing and context." In: Masson-Delmotte, V., P. Zhai, H.-O. Portner, D. Roberts, J. Skea, P. R. Shukla, et al. (eds.), *Global Warming of 1.5°C. An IPCC Special Report on the Impacts of Global Warming of 1.5°C above Pre-Industrial Levels and Related Global Greenhouse Gas Emission Pathways, in the Context of Strengthening the Global Response to the Threat of Climate Change, Sustainable Development, and Efforts to Eradicate Poverty*, 2018.

Archer, D., Eby, M., Brovkin, V., Ridgwell, A., Cao, L., Mikolajewicz, U., Caldeira, K., et al. "Atmospheric lifetime of fossil fuel carbon dioxide." *Annual Review of Earth and Planetary Sciences*, 37, no. 1 (2009): 117–134.

Armstrong McKay, D. I. "Two decades of climate tipping points research: Progress and outlook." *Dialogues on Climate Change*, 1, no. 1 (2024): 30–38. https://doi.org/10.1177/29768659241293272 (Original work published 2024)

Armstrong McKay, D. I., Staal, A., Abrams, J. F., Winkelmann, R., Sakschewski, B., Loriani, S., Fetzer, I., Cornell, S.E., Rockström, J., and Lenton, T. M. "Exceeding 1.5 C global warming could trigger multiple climate tipping points." *Science*, 377, no. 6611 (2022): eabn7950.

Arrhenius, S. "XXXI. On the influence of carbonic acid in the air upon the temperature of the ground." *The London, Edinburgh, and Dublin Philosophical Magazine and Journal of Science*, 41, no. 251 (1896): 237–276.

Balsari, Satchit, Dresser, C., and Leaning, J. "Climate change, migration, and civil strife." *Current Environmental Health Reports*, 7, no. 4 (2020): 404–414.

Bhatia, Kieran, Baker, A., Yang, W., Vecchi, G., Knutson, T., Murakami, H., Kossin, J. et al. "A potential explanation for the global increase in tropical cyclone rapid intensification." *Nature communications*, 13, no. 1 (2022): 6626.

Bjerknes, V. (1904). "Das Problem der Wettervorhersage, betrachtet vom Standpunkte der Mechanik und der Physik." [The problem of weather prediction, considered from the viewpoints of mechanics and physics]. *Meteorologische Zeitschrift*, 21(1–7). (translated and edited by Volkem E. and S. Brönnimann–*Meteorol Z*. 18 (2009), 663–667).

Boberg, J. *Liquid Assets: How Demographic Changes and Water Management Policies Affect Freshwater Resources*. Rand Corporation, 2005.

Broecker, W. "The great ocean conveyor: Discovering the trigger for abrupt climate change." (2010): 1–176.

Cane, Mark A. "The evolution of El Niño, past and future." *Earth and Planetary Science Letters*, 230, no. 3–4 (2005): 227–240.

Chiang, F., Mazdiyasni, O., and AghaKouchak, A. "Evidence of anthropogenic impacts on global drought frequency, duration, and intensity." *Nature Communications*, 12, no. 1 (2021): 1–10.

Cioffi, F., Conticello, F., and Lall, U. "Projecting changes in Tanzania rainfall for the 21st century." *International Journal of Climatology*, 36, no. 13 (2016): 4297–4314.

Cioffi, F., Conticello, F., Lall, U., Marotta, L., and Telesca, V.. "Large scale climate and rainfall seasonality in a Mediterranean area: Insights from a non-homogeneous Markov model applied to the Agro-Pontino plain." *Hydrological Processes*, 31, no. 3 (2017): 668–686.

Cioffi, F., Conticello, F. R., and Lall, U. "Stochastic Scenarios for 21st century rainfall seasonality, daily frequency, and intensity in South Florida." *Journal of Water Resources Planning and Management*, 146, no. 8 (2020): 04020058.

Cioffi, F., De Bonis Trapella, A., Giannini, M., and Lall, U. "A flood risk management model to identify optimal defence policies in coastal areas considering uncertainties in climate projections." *Water*, 14, no. 9(2022): 1481.

Cioffi, F. and Gallerano, F. "Multi-objective analysis of dam release flows in rivers downstream from hydropower reservoirs." *Applied Mathematical Modelling*, 36, no. 7 (2012): 2868–2889.

Cioffi, F., Lall, U., Rus, E., and Krishnamurthy, C. K. B. "Space-time structure of extreme precipitation in Europe over the last century." *International Journal of Climatology*, 35, no. 8 (2015): 1749–1760.

Cioffi, F., Sun, X., Jiang, Q., Attorre, F., Manes, F., Gyampo, M.A., Giannini, M., Shafei, A., Ferroni, F., Micalizzi, K. and Sebastiani, A. "Water resource management for sustainable agriculture: A framework to adapt to a changing climate in Ghana's Tree Crop Sector." In *Proceedings of the 41ST Iahr World Congress* (pp. 22–27), 2025, June.

Cioffi, F., Tieghi, L., Giannini, M., and Pirozzoli, S. "Flash flood prediction in St.Lucia island through a surrogate hydraulic model." *Journal of Applied Water Engineering and Research*, 12, no. 3 (2024): 297–310.

Colman, Robert, and Soden, Brian J. "Water vapor and lapse rate feedbacks in the climate system." *Reviews of Modern Physics*, 93, no. 4 (2021): 045002.

Constantin, P., and Foias, C. *Navier-Stokes Equations*. University of Chicago Press, 1988.

Conticello, Federico, Cioffi, F., Merz, B., and Lall, U. "An event synchronization method to link heavy rainfall events and large-scale atmospheric circulation features." *International Journal of Climatology*, 38, no. 3 (2018): 1421–1437.

Cook, B. I., Mankin, J. S., Marvel, K., Williams, A. P., Smerdon, J. E., and Anchukaitis, K. J. "Twenty-first century drought projections in the CMIP6 forcing scenarios." *Earth's Future*, 8 (2020): e2019EF001461. https://doi.org/10.1029/2019EF001461

Cosgrove, William J., and Loucks, Daniel P. "Water management: Current and future challenges and research directions." *Water Resources Research*, 51, no. 6 (2015): 4823–4839.

Crutzen, Paul J. "The "anthropocene"." *Journal de Physique IV (Proceedings)*, 12, no. 10 (2002): 1–5.

Dalton, John. *Experiments and Observations to Determine Whether the Quantity of Rain and Dew Is Equal to the Quantity of Water Carried off by the Rivers and Raised by Evaporation: With an Enquiry into the Origin of Springs*. R. & W. Dean, 1802.

Deb, K., Agrawal, S., Pratap, A., and Meyarivan, T. (2002). "A fast elitist non-dominated sorting genetic algorithm for multi-objective optimization: NSGA-II." *Parallel Problem Solving from Nature PPSN VI*, 849–858. https://doi.org/10.1007/3-540-45356-3_83

Deutloff, J., Held, H., and Lenton, T. M. "High probability of triggering climate tipping points under current policies modestly amplified by Amazon dieback and permafrost thaw." *Earth System Dynamics*, 16, no. 2 (2025): 565–583.

Disse, M., Johnson, T. G., Leandro, J., and Hartmann, T. "Exploring the relation between flood risk management and flood resilience." *Water Security*, 9 (2020): 100059

Donat, Markus G., Lowry, Andrew L., Alexander, Lisa V., O'Gorman, Paul A., and Maher, N. "More extreme precipitation in the world's dry and wet regions." *Nature Climate Change*, 6, no. 5 (2016): 508–513.

Dong, B., Ullrich, P., Lee, J., Gleckler, P., Chang, K., and O'Brien, T. A. "A new metrics framework for quantifying and intercomparing atmospheric rivers in observations, reanalyses, and climate models." *Geoscientific Model Development*, 18 (2025): 961–976. https://doi.org/10.5194/gmd-18-961-2025

Doss-Gollin, J., Farnham, D. J., Steinschneider, S., and Lall, U. "Robust adaptation to multiscale climate variability." *Earth's Future*, 7, no. 7 (2019): 734–747.

Duffy, C. J. "The terrestrial hydrologic cycle: an historical sense of balance." *Wiley Interdisciplinary Reviews: Water*, 4, no. 4 (2017): e1216.

Dussaillant, I., Hugonnet, R., Huss, M., Berthier, E., Bannwart, J., Paul, F., and Zemp, M. "Annual mass change of the world's glaciers from 1976 to 2024 by temporal downscaling of satellite data with in situ observations." *Earth System Science Data*, 17, no. 5 (2025): 1977–2006.

Elgendy, M., Hassini, S., and Coulibaly, P. (2024). "Review of climate change adaptation strategies in water management." *Journal of Hydrologic Engineering*, 29, no. 1: 03123001.

Emanuel, K. "Response of global tropical cyclone activity to increasing CO_2: Results from downscaling CMIP6 models." *Journal of Climate*, 34, no. 1 (2021): 57–70.

Fan, J., Meng, J., Ludescher, J., Chen, X., Ashkenazy, Y., Kurths, J., Havlin, J., and Schellnhuber, H. J. "Statistical physics approaches to the complex Earth system." *Physics Reports*, 896 (2021): 1–84.

Folke, C., Polasky, S., Rockström, J., Galaz, V., Westley, F., Lamont, M., Scheffer, M., et al. "Our future in the Anthropocene biosphere." *Ambio* (2021): 1–36.

Friedlingstein, P., Artaxo, P., Gallego-Sala, A.V., Jia, G., Jones, C.D., Kawamiya, M., Loisel, J., Loutre, M.F., Rehfeld, K., Rovere, A. and Smith, C., "Earth system responses to different levels of greenhouse gas emissions mitigation." *Frontiers in Climate*, 6 (2024): 1480208.

Friedlingstein, P., Le Quéré, C., O'Sullivan, M., Hauck, J., Landschützer, P., Luijkx, I.T., Li, H., van der Woude, A., Schwingshackl, C., Pongratz, J., and Regnier, P. "Emerging climate impact on carbon sinks in a consolidated carbon budget." *Nature*, (2025a): 1–3.

Friedlingstein, P., O'Sullivan, M., Jones, M. W., Andrew, R. M., Bakker, D. C., Hauck, J., Landschützer, P., Le Quéré, C., Li, H., Luijkx, I. T., and Peters, G. P. "Global carbon budget 2025." *Earth System Science Data Discussions*, 2025 (2025b): 1–139.

Ganopolski, A., and Rahmstorf, S. "Rapid changes of glacial climate simulated in a coupled climate model." *Nature*, 409, no. 6817 (2001): 153–158.

Gerya, T. V., Stern, R. J., Turner, F., et al. "The long-term survival of human civilization: A science-based paradigm." *Habitable Planet*, 2, no. 1, (2026): 1–13. https://doi.org/10.63335/j.hp.2025.0022

Ghil, M., and Lucarini, V. "The physics of climate variability and climate change." *Reviews of Modern Physics*, 92, no. 3 (2020): 035002.

Gleeson, T., Wang-Erlandsson, L., Porkka, M., Zipper, S. C., Jaramillo, F., Gerten, D., et al "Illuminating water cycle modifications and Earth system resilience in the Anthropocene." *Water Resources Research*, 56(2020a): e2019WR024957. https://doi.org/10.1029/2019WR024957

Gleeson, T., Wang-Erlandsson, L., Zipper, S. C., Porkka, M., Jaramillo, F., Gerten, D., Fetzer, I. et al. "The water planetary boundary: Interrogation and revision." *One Earth*, 2, no. 3 (2020b): 223–234.

Gleick, P. H. "Global freshwater resources: soft-path solutions for the 21st century." *Science*, 302, no. 5650 (2003a): 1524–1528.

Gleick, P. H. "Water use." *Annual Review of Environment and Resources*, 28, no. 1 (2003b): 275–314.

Glikson, A. "The lungs of the Earth: Review of the carbon cycle and mass extinction of species." *Energy Procedia*, 146 (2018): 3–11.

Golding, N., and Betts, R. "Fire risk in Amazonia due to climate change in the HadCM3 climate model: Potential interactions with deforestation." *Global Biogeochemical Cycles*, 22, no. 4 (2008).

Granata, F., and Di Nunno, F. "Pathways for hydrological resilience: Strategies for adaptation in a changing climate." *Earth Systems and Environment*, 10 (2026): 203–231. https://doi.org/10.1007/s41748-024-00567-x

He, C., Liu, Z., Jianguo Wu, Pan, X., Fang, Z., Li, J., and Bryan, B. A.. "Future global urban water scarcity and potential solutions." *Nature Communications*, 12, no. 1 (2021): 4667.

Hegerl, G. C., Brönnimann, S., Cowan, T., Friedman, A. R., Hawkins, E., Iles, C., Müller, W., Schurer, A., and Undorf, S., Causes of climate change over the historical record. *Environmental Research Letters*, 14, no. 12 (2019): 123006.

Hermans, T. H. J., et al. "Projecting global mean sea-level change using CMIP6 models." *Geophysical Research Letters*, 48, no. 5 (2021): e2020GL092064.

Hoegh-Guldberg, O., Jacob, D., Taylor, M., Bindi, M., Brown, S., Camilloni, I., Diedhiou, A., Djalante, R., Ebi, K. L., Engelbrecht, F., Guiot, J., Hijioka, Y., Mehrotra, S., Payne, A., Seneviratne, S. I., Thomas, A., Warren, R., and Zhou, G. "Impacts of 1.5°C global warming on natural and human systems." In: Masson-Delmotte, V., P. Zhai, H.-O. Pörtner, D. Roberts, J. Skea, P. R. Shukla, A. Pirani, W. Moufouma-Okia, C. Péan, R. Pidcock, S. Connors, J. B. R. Matthews, Y. Chen, X. Zhou, M. I. Gomis, E. Lonnoy, T. Maycock, M. Tignor, and T. Waterfield (eds.), *Global Warming of 1.5°C. An IPCC Special Report on the Impacts of Global Warming of 1.5°C Above Pre-industrial Levels and Related Global Greenhouse Gas Emission Pathways, in the Context of Strengthening the Global Response to the Threat of Climate Change, Sustainable Development, and Efforts to Eradicate Poverty*. Cambridge University Press, 2018, pp. 175–312. https://doi.org/10.1017/9781009157940.005

Holt, J., Harle, J., Proctor, R., Michel, S., Ashworth, M., Batstone, C., Allen, I., et al. "Modelling the global coastal ocean." *Philosophical Transactions of the Royal Society A: Mathematical, Physical and Engineering Sciences*, 367, no. 1890 (2009): 939–951.

Houghton, J. T., Giggs, M. F. L. G., David, J., and Maskell, K. "An introduction to simple climate models, box model." In McGuffie, K. and A. Henderson-Sellers (eds.), *Climate Modelling Primer Kendal McGuffie, Ann Henderson-Sellers*. West Sussex: Wiley-Blackwell, 2014.

Hugonnet, R., McNabb, R., Berthier, E., Menounos, B, Nuth, C., Girod, L., Farinotti, D., et al. "Accelerated global glacier mass loss in the early twenty-first century." *Nature*, 592, no. 7856 (2021): 726–731.

Intergovernmental Panel on Climate Change (IPCC). (2021). "Climate change 2021: The physical science basis." In *Contribution of Working Group I to the Sixth*

Assessment Report of the Intergovernmental Panel on Climate Change (Table 2.3, p. 2–44). Cambridge University Press. https://doi.org/10.1017/9781009157896

Janetschek, H., Brandi, C., Dzebo, A., and Hackmann, B.. "The 2030 Agenda and the Paris Agreement: voluntary contributions towards thematic policy coherence." *Climate Policy*, 20, no. 4 (2020): 430–442.

Jongman, B. "Effective adaptation to rising flood risk." *Nature communications*, 9, no. 1 (2018): 1–3.

Karamperidou, C., Cioffi, F., and Lall, U.. "Surface temperature gradients as diagnostic indicators of midlatitude circulation dynamics." *Journal of Climate*, 25, no. 12 (2012): 4154–4171

Kaufhold, C., Willeit, M., Liu, B., and Ganopolski, A.. "Assessing the lifetime of anthropogenic CO 2 and its sensitivity to different carbon cycle processes." *Biogeosciences*, 22, no. 12 (2025): 2767–2801.

Keel, T. "Shifting jet streams in a changing climate." PhD diss., UCL (University College London), 2025.

Kim, Y.-H., et al. "Evaluation of the CMIP6 multi-model ensemble for climate extreme indices." *Weather and Climate Extremes*, 29 (2020): 100269.

Kwadijk, J. C. J., Haasnoot, M., Mulder, J. P. M., Hoogvliet, M. M. C., Jeuken, A. B. M., van der Krogt, R. A. A., ... and de Wit, M. J. M. "Using adaptation tipping points to prepare for climate change and sea level rise: A case study in the Netherlands." *Wiley Interdisciplinary Reviews: Climate Change*, 1, no. 5(2010): 729–740. https://doi.org/10.1002/wcc.64

Lenton, T. M., Held, H., Kriegler, E., Hall, J. W., Lucht, W., Rahmstorf, S., and Schellnhuber, H. J.. "Tipping elements in the Earth's climate system." *Proceedings of the National Academy of Sciences*, 105, no. 6 (2008): 1786–1793.

Lenton, T. M., Rockström, J, Gaffney, O., Rahmstorf, S., Richardson, K., Steffen, W., and Schellnhuber, H. J.. "Climate tipping points, too risky to bet against." (2019): 592–595.

Lenton, T. M., and Williams, H. T.P.. "On the origin of planetary-scale tipping points." *Trends in Ecology & Evolution*, 28, no. 7 (2013): 380–382.

Little, C. M., and Urban, N. M. "CMIP5 temperature biases and 21st century warming around the Antarctic coast." *Annals of Glaciology*, 57, no. 73(2016): 69–78. https://doi.org/10.1017/aog.2016.25

Liu, Y., Engel, B. A., Flanagan, D. C., Gitau, M. W., McMillan, S. K., and Chaubey, I. "A review on effectiveness of best management practices in improving hydrology and water quality: Needs and opportunities." *Science of the Total Environment*, 601 (2017): 580–593.

Lorenz, E. N. "Deterministic nonperiodic flow." *Journal of Atmospheric Sciences*, 20, no. 2 (1963): 130–141.

Lorenz, E. N. "Irregularity: A fundamental property of the atmosphere." *Tellus A*, 36, no. 2 (1984): 98–110.

Lorenz, E. N. "Predictability: A problem partly solved." *Proceedings Seminar on Predictability*, 1, no. 1 (1996).

Loulergue, L., Schilt, A., Spahni, R., Masson-Delmotte, V, Blunier, T., Lemieux, B., Barnola, J.-M., Raynaud, D., Stocker, T. F., and Chappellaz, J. "Orbital and millennial-scale features of atmospheric CH_4 over the past 800,000 years." *Nature*, 453, no. 7193 (2008): 383–386.

Lucarini, V.. "Modeling complexity: The case of climate science." In *Models, simulations, and the reduction of complexity* (pp. 229–254). De Gruyter, 2013.

Lynch, P.. "The origins of computer weather prediction and climate modeling." *Journal of Computational Physics*, 227, no. 7 (2008): 3431–3444.

Manabe, S., Smagorinsky, J., and Strickler, R. F. "Simulated climatology of a general circulation model with a hydrologic cycle." *Monthly Weather Review*, 93, no. 12 (1965): 769–798.

Manabe, S. and Stouffer, R. J. "The role of thermohaline circulation in climate." *Tellus B*, 51, no. 1 (1999): 91–109.

Manabe, S., and Wetherald, R. T. "The effects of doubling the CO_2 concentration on the climate of a general circulation model." *Journal of Atmospheric Sciences*, 32, no. 1 (1975): 3–15.

Matthews, H. D., Gillett, N. P., Stott, P. A., and Zickfeld, K. "The proportionality of global warming to cumulative carbon emissions." *Nature*, 459, no. 7248 (2009): 829–832

Merz, B., Aerts, J. C. J. H., Arnbjerg-Nielsen, K., Baldi, M., Becker, A., Bichet, A., Blöschl, G., Bouwer, L. M., Brauer, A., Cioffi, F., and Delgado, J. M. "Floods and climate: Emerging perspectives for flood risk assessment and management." *Natural Hazards and Earth System Sciences*, 14(7) (2014): 1921–1942.

Morice, C. P., Kennedy, J. J., Rayner, N. A., Winn, J. P., Hogan, E. Killick, R. E. Dunn, R. J. H. Osborn, T. J. Jones, P. D., and Simpson, I. R. "An Updated Assessment of Near-Surface Temperature Change From 1850: The HadCRUT5 Data Set." *Journal of Geophysical Research: Atmospheres*, 126, no. 3 (2021): e2019JD032361.

Muis, S., Güneralp, B., Jongman, B., Aerts, J. C. J. H., and Ward, P. J. "Flood risk and adaptation strategies under climate change and urban expansion: A probabilistic analysis using global data." *Science of the Total Environment*, 538(2015): 445–457. https://doi.org/10.1016/j.scitotenv.2015.08.068

Myhre, G., Shindell, D., Bréon, F.-M., Collins, W., Fuglestvedt, J., Huang, J., D. Koch, Lamarque, J.-F., Lee, D., Mendoza, B., Nakajima, T., Robock, A., Stephens, G., Takemura, T., and Zhang, H. "Anthropogenic and natural radiative forcing." In Stocker, T. F., D. Qin, G.-K. Plattner, M. Tignor, S. K. Allen, J. Boschung, A. Nauels, Y. Xia, V. Bex and P. M. Midgley (eds.), *Climate Change 2013: The Physical Science Basis. Contribution of Working Group I to the Fifth Assessment Report of the Intergovernmental Panel on Climate Change*. Cambridge and New York: Cambridge University Press, 2013.

Nace, R. L. "The hydrological cycle: Historical evolution of the concept." *Water International*, 1, no. 1 (1975): 15–21.

Nagata, K., Shoji, I., Arima, T., Otsuka, T., Kato, K., Matsubayashi, M., and Omura, M. "Practicality of integrated water resources management (IWRM) in different contexts." *International Journal of Water Resources Development*, 38, no. 5(2022): 897–919. https://doi.org/10.1080/07900627.2021.1921709

Nese, J. M., and Grenci, L. M. *A World of Weather: Fundamentals of Meteorology*, 5th edn. Kendall Hunt Publishing, 2011.

Neukom, R., Steiger, N., Gómez-Navarro, J. J., Wang, J., and Werner, J. P. "No evidence for globally coherent warm and cold periods over the preindustrial common era." *Nature*, 571, no. 7766 (2019): 550–554.

Obe, C.B. "The mathematics of climate change." 2018. http://content.gresham.ac.uk.s3.amazonaws.com/data/binary/2888/2018-11-13_ChrisBudd_MathsOfClimateChange-T.pdf

O'Neill, B. C., Kriegler, E., Ebi, K. L., Kemp-Benedict, E., Riahi, K., Rothman, D. S., ... and Solecki, W. "The roads ahead: Narratives for shared socioeconomic pathways describing world futures in the 21st century." *Global Environmental Change*, 42(2017): 169–180. https://doi.org/10.1016/j.gloenvcha.2015.01.004

O'Neill, B. C., Tebaldi, C., Vuuren, D. P. V., Eyring, V., Friedlingstein, P., Hurtt, G.,… and Sanderson, B. M. "The scenario model intercomparison project (ScenarioMIP) for CMIP6." *Geoscientific Model Development*, 9, no. 9(2016): 3461–3482

Oki, T., and Kanae, S. "Global hydrological cycles and world water resources." *Science*, 313, no. 5790(2006): 1068–1072.

Orbe, C., Rind, D., Miller, R. L., Nazarenko, L. S., Romanou, A., Jonas, J., Russell, G. L., Kelley, M., and Schmidt, G. A.. "Atmospheric response to a collapse of the North Atlantic circulation under a mid-range future climate scenario: A regime shift in northern hemisphere dynamics." *Journal of Climate*, 36, no. 19 (2023): 6669–6693.

Pacher, B., Lalic, B., and Eitzinger, J. Marta, A. D., Orlandini, S., and Sremac, A. F. *Agricultural Meteorology and Climatology*. Firenze University Press, 2018.

Papalexiou, S. M., and Mamalakis, A. "Machine unlearning: bias correction in neural network downscaled storms." *Journal of Hydrology*, 665 (2026).

Parker, W. S. "Predicting weather and climate: Uncertainty, ensembles and probability." *Studies in history and Philosophy of Science Part B: Studies in History and Philosophy of Modern Physics*, 41, no. 3 (2010): 263–272.

Penduff, T., Juza, M., Barnier, B., Zika, J., Dewar, W. K., Treguier, A. M., … Audiffren, N.. "Sea level expression of intrinsic and forced ocean variabilities at interannual time scales." *Journal of Climate*, 24, no. 21(2011): 5652–5670. https://doi.org/10.1175/JCLI-D-11-00077.1

Phillips, N. A. "The general circulation of the atmosphere: A numerical experiment." *Quarterly Journal of the Royal Meteorological Society*, 82, no. 352 (1956): 123–164.

Pierrehumbert, R. T. "Infrared radiation and planetary temperature." In *AIP Conference Proceedings*, vol. 1401, no. 1, pp. 232–244. American Institute of Physics, 2011.

Provenzale, A.. "Climate models." *Rendiconti Lincei*, 25, no. 1 (2014): 49–58.

Rahmstorf, S. "Bifurcations of the Atlantic thermohaline circulation in response to changes in the hydrological cycle." *Nature*, 378, no. 6553 (1995): 145–149.

Ranzi, R., Bae, D.-H., Nalder, G., Losada, I. J., Kobayashi, K., Teegavarapu, R., and Silva, T. "WATER MONOGRAPHS. guidance for water engineering in a Changing Climate." (2025)

Regnier, P., Friedlingstein, P., Ciais, P., Mackenzie, F. T., Gruber, N., Janssens, I. A., Laruelle, G. G., et al. "Anthropogenic perturbation of the carbon fluxes from land to ocean." *Nature Geoscience*, 6, no. 8 (2013): 597–607.

Rehfeld, K., and Ziegler, E. *Future Climate in the Context of the Earth System Evolution over the last 66 Million Years*. figshare, 2023. https://doi.org/10.6084/m9.figshare.22732775.v1

Riahi, K., Van Vuuren, D.P., Kriegler, E., Edmonds, J., O'Neill, B.C., Fujimori, S., Bauer, N., Calvin, K., Dellink, R., Fricko, O., and Lutz, W. "The shared socioeconomic pathways and their energy, land use, and greenhouse gas emissions implications: An overview." *Global Environmental Change*, 42 (2017): 153–168.

Rial, J. A., Pielke, R. A., Beniston, M., Claussen, M., Canadell, J., Cox, P., Held, H., et al. "Nonlinearities, feedbacks and critical thresholds within the Earth's climate system." *Climatic Change*, 65, no. 1 (2004): 11–38.

Richardson, K., Steffen, W., Lucht, W., Bendtsen, J., Cornell, S. E., Donges, J. F., Drüke, M., et al. "Earth beyond six of nine planetary boundaries." *Science Advances*, 9, no. 37 (2023): eadh2458.

Richardson, L. F. *Weather Prediction by Numerical Process*. Cambridge University Press, 1922.

Rockström, J., Donges, J. F., Fetzer, I., et al. "Planetary boundaries guide humanity's future on Earth." *Nature Reviews Earth & Environment*, 5 (2024): 773–788.

Rockstrom, J., Steffen, W., Noone, K., Persson, A., Chapin, F. S. III, et al.. "A safe operating space for humanity." *Nature*, 461, no. 7263(2009a):472–752.

Rogelj, J., Shindell, D., Jiang, K., Fifita, S., Forster, P., Ginzburg, V., Handa, C., Kheshgi, H., Kobayashi, S., Kriegler, E., Mundaca, L., Séférian, R., and Vilarino, M. V. "Mitigation pathways compatible with 1.5°C in the context of sustainable development." In: Masson-Delmotte, V., P. Zhai, H.-O. Pörtner, D. Roberts, J. Skea, P.R. Shukla, A. Pirani, W. Moufouma-Okia, C. Péan, R. Pidcock, S. Connors, J.B.R. Matthews, Y. Chen, X. Zhou, M.I. Gomis, E. Lonnoy, T. Maycock, M. Tignor, and T. Waterfield (eds.), *Global Warming of 1.5°C. An IPCC Special Report on the Impacts of Global Warming of 1.5°C above Pre-Industrial Levels and Related Global Greenhouse Gas Emission Pathways, in the Context of Strengthening the Global Response to the Threat of Climate Change, Sustainable Development, and Efforts to Eradicate Poverty*. In Press, 2018.

Roy, J., Tschakert, P., Waisman, H., Halim, S. Abdul, Antwi-Agyei, P., Dasgupta, P., Hayward, B., Kanninen, M., Liverman, D., Okereke, C., Pinho, P. F., Riahi, K., and Rodriguez, A. G. S. "Sustainable development, poverty eradication and reducing inequalities." In: Masson-Delmotte, V., P. Zhai, H.-O. Pörtner, D. Roberts, J. Skea, P.R. Shukla, A. Pirani, W. Moufouma-Okia, C. Péan, R. Pidcock, S. Connors, J. B. R. Matthews, Y. Chen, X. Zhou, M. I. Gomis, E. Lonnoy, T. Maycock, M. Tignor, and T. Waterfield (eds.), *Global Warming of 1.5°C. An IPCC Special Report on the Impacts of Global Warming of 1.5°C above Pre-Industrial Levels and Related Global Greenhouse Gas Emission Pathways, in the Context of Strengthening the Global Response to the Threat of Climate Change, Sustainable Development, and Efforts to Eradicate Poverty* (pp. 445–538). Cambridge and New York: Cambridge University Press, 2018. https://doi.org/10.1017/9781009157940.007

Schellnhuber, H. J. "Tipping elements in the Earth System." *Proceedings of the National Academy of Sciences*, 106, no. 49 (2009): 20561–20563.

Scoccimarro, E., and Gualdi, S. "Heavy daily precipitation events in the CMIP6 worst-case scenario: Projected twenty-first-century changes." *Journal of Climate*, 33, no. 17 (2020): 7631–7642.

Shafei, A., and Cioffi, F. *High-Resolution Early Warning Systems Using DL: Part I-Elevation-Integrated Temperature and Precipitation SRGAN Downscaling (E-TEPS)*. Preprints.org, 2024a.

Shafei, A., and Cioffi, F. *High-Resolution Early Warning Systems Using DL: Part II-Combining FourCastNet and E-TEPS for High-Resolution Climate Forecasting*. Preprint.org, 2024b.

Shelton, M. L. *Hydroclimatology: Perspectives and Applications*. Cambridge University Press, 2009.

Spinoni, J., Barbosa, P., De Jager, A., McCormick, N., Naumann, G., Vogt, J. V., Magni, D., Masante, D., and Mazzeschi, M. "A new global database of meteorological drought events from 1951 to 2016." *Journal of Hydrology: Regional Studies*, 22 (2019): 100593.

Steffen, W., Rockström, J., Richardson, K., Lenton, T. M., Folke, C., Liverman, D., Summerhayes, C. P., et al. "Trajectories of the Earth System in the Anthropocene." *Proceedings of the National Academy of Sciences*, 115, no. 33 (2018): 8252–8259.

Stevens, B., and Bony, S. "Water in the atmosphere." *Physics Today*, 66, no. 6 (2013): 29–34.

Stommel, H. "Thermohaline convection with two stable regimes of flow." *Tellus*, 13, no. 2 (1961): 224–230.

Tapiador, F. J., Navarro, A., Moreno, R., Sánchez, J. L., and García-Ortega, E. "Regional climate models: 30 years of dynamical downscaling." *Atmospheric Research*, 235 (2020): 104785.

Tebaldi, C., and Friedlingstein, P. "Delayed detection of climate mitigation benefits due to climate inertia and variability." *Proceedings of the National Academy of Sciences*, 110, no. 43 (2013): 17229–17234.

Tebaldi, C., et al. "Climate model projections from the scenario model intercomparison project (Scenario MIP) of CMIP6." *Earth System Dynamics*, 12, no. 1 (2021): 253–293.

Teutschbein, C., and Seibert, J. "Bias correction of regional climate model simulations for hydrological climate-change impact studies: Review and evaluation of different methods." *Journal of Hydrology* 456 (2012): 12–29.

Touzé-Peiffer, L., Barberousse, A., and Le Treut, H. "The coupled model intercomparison project: History, uses, and structural effects on climate research. WIREs" *Clim Change*, 11(2020): e648. https://doi.org/10.1002/wcc.648

Trenberth, K. E., Fasullo, J. T., and Kiehl, J. "Earth's global energy budget." *Bulletin of the American Meteorological Society*, 90, no. 3 (2009): 311–324.

Turner, A. J., Frankenberg, C., and Kort, E. A. "Interpreting contemporary trends in atmospheric methane." *Proceedings of the National Academy of Sciences*, 116, no. 8 (2019): 2805–2813.

Tyndall, J. "XXIII. On the absorption and radiation of heat by gases and vapours, and on the physical connexion of radiation, absorption, and conduction. The bakerian lecture." *The London, Edinburgh, and Dublin Philosophical Magazine and Journal of Science*, 22, no. 146 (1861): 169–194.

Van der Ent, R. J., Savenije, H. H. G., Schaefli, B., and Steele-Dunne, S. C. "Origin and fate of atmospheric moisture over continents." *Water Resources Research*, 46, no. 9 (2010).

Van der Ent, R. J., and Tuinenburg, O. "A.: The residence time of water in the atmosphere revisited." *Hydrology and Earth System Sciences*, 21 (2017): 779–790. https://doi.org/10.5194/hess-21-779-2017

Van Dijk, A., Beck, H., Boergens, E., Dorigo, W., Edirisinghe, C., Forootan, E., Guo, E., et al. *Global Water Monitor 2024*. Summary Report. Global Water Monitor Consortium, 2025.

Vardon, M. J., Le, T. H. L., Martinez-Lagunes, R., Pule, O. B., Schenau, S., May, S., and Quentin Grafton, R. "Accounting for water: A global review and indicators of best practice for improved water governance." *Ecological Economics*, 227 (2025): 108396.

Wang, B., Jin, C., and Liu, J. "Understanding future change of global monsoons projected by CMIP6 models." *Journal of Climate*, 33, no. 15 (2020): 6471–6489.

Wang, C. "Three-ocean interactions and climate variability: A review and perspective." *Climate Dynamics*, 53, no. 7 (2019): 5119–5136.

Wang-Erlandsson, L., Tobian, A., Van der Ent, R. J., Fetzer, I., te Wierik, S., Porkka, M.,… and Rockström, J.. "A planetary boundary for green water." *Nature Reviews Earth & Environment*, 3, no. 6(2022): 380–392.

Ward, P. "The geography of future water challenges." (2018).

Ward, P. J., de Ruiter, M. C., Mård, J., Schröter, K., Van Loon, A., Veldkamp, T., von Uexkull, N., et al. "The need to integrate flood and drought disaster risk reduction strategies." *Water Security*, 11 (2020): 100070.

Weart, S. "The development of general circulation models of climate." *Studies in History and Philosophy of Science Part B: Studies in History and Philosophy of Modern Physics*, 41, no. 3 (2010): 208–217.

Woodward, M., Kapelan, Z., and Gouldby, B. "Adaptive flood risk management under climate change uncertainty using real options and optimization." *Risk Analysis*, 34, no. 1(2014): 75–92. https://doi.org/10.1111/risa.12088

Wunderling, N., von der Heydt, A. S., Aksenov, Y., Barker, S., Bastiaansen, R., Brovkin, V., Brunetti, M., Couplet, V., Kleinen, T., Lear, C. H. and Lohmann, J. "Climate tipping point interactions and cascades: A review." *Earth System Dynamics*, 15, no. 1 (2024): 41–74.

Yuan, X., Wang, Y., Ji, P., Wu, P., Sheffield, J., and Otkin, J. A. "A global transition to flash droughts under climate change." *Science*, 380 (2023): 187–191.

Zipper, S. C., et al. "Integrating the water planetary boundary with water management from local to global scales." *Earth's Future*, 8, no. 2 (2020): e2019EF001377.

Chapter 5

Ecology and ecosystem systems: Relevant SDGs: 15, 3, 11, 13 and 14

Fausto Manes

5.1 THE COMPLEXITY OF ECOLOGICAL SYSTEMS

5.1.1 Introduction

Ecology studies the relationships among living organisms, as well as between organisms and their environment. It deals with plant, animal, fungal, and microbial communities, the flow of energy, the cycling of matter, and the interactions with environmental resources and conditions. According to the definition given by G. E. Hutchinson in 1958, ecology is "the science of the universe, the study of how the world works."

The relationships among living organisms may be trophic, that is, concerning the feeding interactions among different species within an ecosystem. These relationships begin with the first trophic level, composed of autotrophs, which are able to produce complex organic substances through photosynthesis, deriving their energy from sunlight and using simple inorganic compounds. The subsequent trophic level is occupied by heterotrophs, that is, animals that ingest organic matter; these include herbivores, carnivores, and decomposer organisms (bacteria, fungi, and detritivores), the latter feeding on dead organic matter or detritus. The distinction between autotrophs and heterotrophs, as well as their trophic interactions, varies in complexity depending on the type of ecosystem considered—be it a grassland or forest for terrestrial ecosystems, a river or lake for freshwater systems, or marine ecosystems.

Ecology also investigates the relationships between organisms and the physicochemical characteristics of the environment. Particularly significant are the distribution and abundance of a plant or animal species, or of groups of individuals belonging to that species, within an area where suitable conditions exist—conditions that ensure appropriate habitats consistent with the ecological requirements of the species. It is also important to analyze the geographical distribution (range) of a species. For example, within the boundaries of their respective ranges, one may find the Turkey oak (*Quercus cerris*), a deciduous plant species, or the holm oak (*Quercus ilex*), an evergreen species, if the environmental conditions exist to allow survival and reproduction.

 DOI: 10.1201/9781003595700-6

From these brief notes, it becomes evident that ecology is a biological discipline requiring knowledge from other fields such as chemistry, geology, geography, climatology, economics, and the social sciences. The analysis of ecological systems is therefore inherently complex.

The terms **ecology** (from the Greek *oikos*, meaning "house" seen as a "habitat," and *logos*, meaning "study" or "discourse") and **economy** (from *oikos* and *nomos*, "laws" or "management of the house") appear semantically similar, despite referring to distinct sciences that are sometimes perceived as antithetical. The term ecology was first introduced by H. Reiter in 1865, but it was E. H. Haeckel in 1869 who defined its scope as "the science that studies the economy of nature and the relations of animals to their inorganic and organic environment." The discipline gained recognition during the second half of the 20th century. E. P. Odum's 1973 definition, describing ecology as "the study of the structure and function of ecosystems," remains highly relevant and aligns with the main trends of international scientific research in the field. Odum also introduced the concepts of **structure** and **function**, applicable to all levels of biological organization—from the macromolecular scale to organisms, communities, ecosystems, and landscapes. The latter are conceived as mosaics of ecosystems (forests, grasslands, croplands, and urban areas) within a given territory.

Today, when referring to global change, we mean ecosystem processes that have been in some way altered by human activities. Thus, when studying ecosystem functioning, it is essential to analyze the structural characteristics underpinning these processes.

5.1.2 The fundamental concepts of ecology

Several terms and concepts underpin the evolutionary path of ecological studies. One such term is **autoecology**, which concerns the relationships within a species or between individuals of a population (i.e., a group of individuals belonging to the same species) and the abiotic factors of their environment. Autoecology thus consists of the study of a species and the set of factors that determine its life conditions.

Synecology, on the other hand, examines how conspecific and heterospecific organisms form populations and communities, how these change over time, and how they interact with the abiotic environment. In other words, synecology studies systems of populations belonging to different animal, plant, and fungal species that are linked by trophic relationships, forming communities or biocoenoses.

The importance of these studies lies in the fact that ecological systems, through their functional processes, generate goods and ecosystem services essential for humanity (Ws. 1, IPBES).

An ecosystem is an environment (such as a river, lake, grassland, forest, or even a city) that develops over space and time with its own identity. It is characterized by the network of relationships (e.g., food web) established among mainly green plants and algae (producers), animals (consumers), fungi, bacteria (major decomposers), and detritivores (e.g., earthworms in terrestrial ecosystems and mollusks in aquatic ecosystems). An ecosystem can therefore be regarded as an entity composed of the **biocoenosis** (the living community) and the **biotope** (the surrounding physical environment).

Taken as a whole, the ecosystem is a functional unit that includes biocoenoses and abiotic environmental components and that possesses its own structure, functioning, and temporal development. Structure, for instance, may be represented by the vertical stratification of a tropical rainforest. Within such a forest, one may observe an initial layer composed mainly of mosses, followed by an herbaceous layer, then a layer of shrubs and small trees. Proceeding in height, the canopy reveals dominant trees and, above them, emergent trees that may reach heights of 30–40 meters.

This dense canopy cover, characteristic of rainforests, constitutes a significant structural and functional feature. It can be described using the *Leaf Area Index* (LAI), defined as the ratio between the total leaf surface area in the canopy of a plant (m^2) and the ground surface area onto which those leaves are projected (m^2). A bare soil has an LAI value of zero, while in a tropical rainforest this index can reach values of 10 or more—that is, 10 or more square meters of leaf surface area are stratified over each square meter of ground. This index is of fundamental importance to analyze processes such as photosynthesis, the vegetation's ability to produce biomass, and leaf transpiration, which plays a crucial role in the hydrological cycle, particularly in climate regulation (see the water cycle, Chapter 4, Section 4.3.2)

Structural complexity concerns not only the above-ground part of the ecosystem but also the below-ground components. For example, the root systems present in the soil develop at various depths, reducing competition among plant species for water and nutrient uptake. Soil itself is a complex system, and vegetation, through its structure and function, influences the biogeochemical cycles (Ws.2, IPCC).

The study of an ecosystem must also consider its temporal dynamics. For example, the primary succession, in a terrestrial environment, over time, evolves from an area initially composed only of bare rock and devoid of life; rocks may be colonized by pioneer organisms such as mosses and lichens. These pioneer species initiate the process of primary succession by enriching the mineral soil with organic matter and contributing to soil formation, alongside geological weathering processes that break down bedrock and minerals, releasing nutrients into circulating soil water. Organic matter begins to accumulate, and the substrate increases in fertility and humus content (Ws. 3, EU Soil Strategy for 2030). Such soil can then support the germination of more demanding plant species, giving rise to a herbaceous community. This community is followed by shrubs, and over decades or centuries, a forest may eventually develop. This stage (once referred to as the *climax*) represents a "mature community": a complex community that constitutes the final stage of a primary succession, consistent with the climatic and edaphic conditions of the site.

This evolutionary process, known as **primary succession**, is characterized by successive stages in which plant, animal, and fungal species dynamically replace one another, forming different communities from the initial stage to maturity. The mature community is therefore the most advanced stage attainable under the given climatic and soil conditions and may consist of a herbaceous community (e.g., an alpine meadow), a shrubland (e.g., Mediterranean scrub), or a woodland (e.g., oak, holm oak, or beech forest), depending, as already specified, on the characteristics of the site.

An example of primary succession can arise from a lava flow. During a volcanic eruption, any forest ecosystems on the affected slopes are destroyed, along with all other forms of life. After the lava solidifies, these slopes remain covered by a mineral substrate without any form of life, including microorganisms. From this state, a primary succession process can begin: if further eruptions do not occur on that slope, within a few centuries, the pre-existing forest may re-establish itself through the evolutionary dynamics of vegetation.

Secondary succession, by contrast, is another dynamic process of vegetation change. It begins on a substrate that has already supported natural biocoenoses or agricultural crops and thus already has a developed soil. Consequently, the development of the mature community occurs more rapidly (over decades rather than centuries). For instance, a field once cultivated and then abandoned quickly hosts herbaceous species, followed by shrubs, and eventually a woodland, if the environmental site conditions are suitable. This process is particularly evident in Italy in the Apennine region, where many lands were abandoned in the past century as rural populations migrated to urban areas or lowlands. The fields previously cultivated and then abandoned today host true forest communities.

It is important to note that in secondary succession, the soil is already more developed than in primary succession, where soil forms progressively through the gradual replacement of species and the slow, complex process of pedogenesis. Similarly, after a fire, the recovery of vegetation in the affected area sees the start of a secondary succession process. In this context, it is useful to remember the EU soil strategy for 2030 (Ws. 3, EU Soil Strategy for 2030).

5.2 NATURAL CAPITAL, ECOSYSTEM SERVICES, BIODIVERSITY, AND NATURE-BASED SOLUTIONS

5.2.1 The ecosystem: Structure and function

The ecosystem model proposed by E. P. Odum schematically illustrates the structure, function, and main ecosystem processes described in Section 5.1. These include a unidirectional flow of energy: solar radiation absorbed by plants (autotrophs) through photosynthesis enables the production of organic matter. This matter, through the *grazing food chain*, leads to the generation of new biomass in heterotrophic organisms (herbivores and carnivores). During this pathway, electromagnetic energy (incident sunlight) is transformed into chemical bond energy stored in the compounds formed by photosynthesis.

Within the ecosystem, one may describe an energy pyramid, in which at each trophic level, part of the energy is dissipated as respiratory heat to sustain the living community (plants, animals, fungi, and microorganisms). Thus, ecosystems are characterized by a unidirectional flow of energy and by the circulation of matter and nutrients through biogeochemical cycles (e.g., the carbon and nitrogen cycles). In parallel, dead organic matter and residues of all organisms—autotrophs and heterotrophs alike—form the *detritus food chain*. Through decomposition processes carried out mainly by bacteria and fungi, nutrients are recycled and become available multiple times within the food web of ecosystems.

The ecosystem, therefore, represents an **open ecological system** in dynamic equilibrium with environmental factors, with its own characteristic structure, functioning, and developmental history. Open systems process inputs and generate outputs under normal conditions; the quantity of outputs depends on the amount of inputs received, and to continue functioning, they require a constant supply of new inputs. It is precisely the continuous flow of solar electromagnetic energy that allows autotrophs, through photosynthesis, to produce the organic matter necessary to sustain the entire community of living beings.

The characteristics and conditions of major ecological systems have been examined through the studies and maps produced within the *Millennium Ecosystem Assessment* (Ws. 4, MEA; Millennium Ecosystem Assessment, 2005), launched by Kofi Annan, UN Secretary-General, in 2000 to evaluate the state of the planet's ecosystems. Maps prepared by researchers show, for example, the distribution of forest ecosystems, aquatic systems, agroecosystems, and urban systems.

In recent years, a specific field of ecology has emerged: urban ecology. More than 50% of the world's population lives in urban areas, a percentage that will continue to rise in the coming decades. This is a crucial consideration for the sustainable management of natural resources and for achieving the targets set by the 17 SDGs of the UN Agenda 2030 (Ws. 5, UN Agenda 2030).

5.2.2 The UN 2030 Agenda and global climate change

The United Nations 2030 Agenda establishes 17 Sustainable Development Goals (SDGs), which are deeply interconnected. Each goal would require specific in-depth analysis: from "Zero Hunger" (SDG 2) to "Good Health and Well-being" (SDG 3)—which is in turn linked to the fight against climate change (SDG 13)—to "Sustainable Cities and Communities"

(SDG 11), "Life Below Water" (SDG 14), and "Life on Land" (SDG 15). Achieving "Affordable and Clean Energy" (SDG 7) is strongly dependent on Industry, Innovation, and Infrastructure" (SDG 9), "Responsible Consumption and Production" (SDG 12), and "Peace, Justice, and Strong Institutions" (SDG 16); all are related to "Quality Education" (SDG 4).

Even with this simple enumeration, it becomes evident that the SDGs are mutually interdependent and all essential for achieving sustainable development. Each goal is articulated into specific targets addressing global challenges to be met by 2030. It is also clear that all SDGs concern, directly or indirectly, the biosphere, and thus the entirety of life on our planet, as well as human activities within different environmental systems (natural ecosystems, agroecosystems, and urban ecosystems). These activities also include economic relations, which are tightly integrated with biosphere processes and represent the foundation of human development and well-being.

A comprehensive discussion of all the SDGs and their targets falls beyond the scope of this text; however, to illustrate their scope, the "Zero Hunger" goal (SDG 2) may serve as an example: approximately 790 million people worldwide (1 in 9 people) are underfed, and the majority of those affected live in developing countries, where 12.9% of the population suffers from malnutrition. By 2030, the key targets include the implementation of sustainable food production systems, the adoption of climate-resilient agricultural practices, and the progressive improvement of soil quality. These measures are crucial to significantly increase agricultural productivity while strengthening adaptation to climate change and to extreme weather events such as droughts and floods.

At the same time, global water pollution is increasing at a rate that exceeds the natural capacity of rivers and lakes to regenerate and self-purify. Despite technological advances that enhance energy efficiency, energy demand in OECD countries is expected to continue rising. SDG 12, "Responsible Consumption and Production," defines crucial targets such as "achieving the sustainable management and efficient use of natural resources, halving per capita global food waste by 2030, and substantially reducing waste generation through prevention, recycling, and reuse."

Within this socioeconomic framework, the challenges related to ecosystem management acquire particular relevance. Ecosystems, which represent **natural capital**, are composed of terrestrial and aquatic systems. These consist of abiotic components—structural habitat features and physical complexity, including resources (light, water, nutrients) and conditions (soil pH, temperature, humidity)—and biotic components, comprising species, their evolutionary history, populations, and communities characterized by ecological and evolutionary processes. A key attribute of biotic environments is **biodiversity**, defined as the variety of life, including genetic diversity, the diversity among individuals, species, populations, communities, ecosystems, and landscapes.

In such a complex context, it is essential to recall the meaning of commonly used terms such as primary production and biomass, and to consider the impacts of anthropogenic activities, including pollution of environmental matrices (air, water, and soil); climate change; and disturbances such as wildfires or intensive grazing. Primary production refers to the assimilation (gross primary production) or accumulation (net primary production) of energy and nutrients by green plants and other autotrophic organisms. Biomass refers to the mass of living matter, usually expressed as dry mass, in the whole or part of an organism, population, or community. It is commonly expressed as mass per unit surface area, a biomass density (g/m^2; t/ha).

Ecology can be divided into basic ecology, which investigates the natural environment and interactions among organisms and with their environment, and applied ecology, which applies the principles of basic ecology to the analysis, monitoring, forecasting, and management of anthropogenic impacts on terrestrial and aquatic ecosystems.

The structural and functional characterization of ecosystems is therefore fundamental for representing dynamic systems composed of plant, animal, fungal, and microbial communities that interact among themselves and with the abiotic environment as a single functional unit.

At present, several critical thresholds for the functioning of biosphere processes have already been surpassed due to anthropogenic pressures. Among the most pressing issues are climate change, environmental pollution, and ocean acidification, as well as alteration of the nitrogen cycle and biodiversity loss (Ws. 6, GLOBAIA). Together with land-use change, these represent some of the most serious environmental challenges.

5.2.3 Ecosystem services, biodiversity, and human well-being

It is essential to introduce the concept of **ecosystem goods and services**, defined as the benefits that human populations derive, directly or indirectly, from ecosystem functions. Goods include, for example, food and raw materials, while services encompass processes such as decomposition and the recycling of organic matter, climate regulation, and pollination. The fundamental importance of understanding ecosystem functioning and biodiversity conservation strategies (plant, animal, fungal, and microbial) emerges, as biodiversity ensures ecosystem functions and, consequently, the processes that provide goods and services crucial for humanity and human well-being (Ws. 7, IPBES Secretariat). Unfortunately, anthropogenic activities are driving profound global changes.

In Daily, 1997, G. C. Daily defined ecosystem services as the conditions and processes through which natural ecosystems and their constituent species sustain and fulfill human life. The Millennium Ecosystem Assessment (Ws.8, MEA) involved over 1,300 researchers worldwide and yielded significant advances in knowledge. The MEA reports, available online, provide comprehensive assessments of how changes in ecosystem health affect human well-being, while also offering a scientific basis for actions aimed at improving conservation and the sustainable use of resources and natural capital, with clear benefits for humanity. This international research effort provided a state-of-the-art overview of ecosystem conditions and services, as well as options for restoring, conserving, or enhancing their sustainable use. Future scenarios were also developed, based on observed ecosystem trends.

The MEA identified four categories of ecosystem services: provisioning, regulating, cultural, and supporting services, the latter being necessary to sustain the others. More recently, the Common International Classification of Ecosystem Services (CICES) (Haines-Young & Potschin, 2018) integrated supporting services within regulating services so that current classifications include provisioning, regulating, and cultural services.

To analyze ecosystem services, it is essential to adopt a broad, transdisciplinary perspective that considers the key relationships between ecosystem services and the components of human well-being. This approach, developed within the *Economics of Ecosystems and Biodiversity* (Ws.9, TEEB) initiative, emphasizes that biodiversity, through ecosystem structure and function, enables the production of services (e.g., flood protection, food, raw materials) that ultimately generate benefits for humans, often quantifiable in monetary terms. In 1997 and later in Costanza et al. (2014), Costanza et al. published a global map assigning monetary values per hectare per year to the major biomes of the planet. Terrestrial ecosystems such as tropical forests and agroecosystems showed particularly high values; temperate ecosystems and coniferous forests (taiga) exhibited intermediate values, while the Sahara Desert, characterized by absent or extremely limited primary production, showed values close to zero.

It is important to distinguish between the *supply* of ecosystem goods and services provided by natural systems and the *demand* for these services within socioeconomic contexts. Both the EU Biodiversity Strategy for 2020 and the more recent 2030 Strategy highlight the interdependence between biodiversity and ecosystem services. Numerous studies have investigated the relationship between biodiversity, ecosystem functioning, and service provision, with experimental evidence consistently indicating that biodiversity has positive effects on most ecosystem services, especially regulating services. While further research is required to fully elucidate these relationships, precautionary principles should be applied in the use and management of biodiversity.

It is possible to describe on a Cartesian plane the relationship between conditions of high and low biodiversity and the level of provision of ecosystem services. At the origin of the x-axis, ecosystems exhibit high biodiversity and naturalness. As land use increases—ranging from initial exploitation to intensive use—conditions deteriorate, biodiversity declines, and ecosystem degradation intensifies. The y-axis shows the level of ecosystem service provision (Braat & de Groot, 2012). Regulating services (e.g., climate regulation, air and water purification, pollination, and other ecosystem processes) are highest in natural ecosystems but decline sharply with biodiversity loss, reaching minimal levels under severe degradation. Provisioning services, limited in pristine ecosystems, reach higher values under sustainable agricultural practices, but decline with overexploitation, eventually leading to reduced or negligible yields. Cultural services, such as recreation, are also greatest in natural environments (except in strictly protected reserves with limited accessibility) but decline alongside biodiversity and ecosystem degradation. Thus, sustainable land use, natural resource management, and biodiversity conservation are key to maintaining the supply of essential goods and services.

Excessive exploitation of biosphere resources has altered the structure and functioning of many ecosystems, often compromising their capacity to provide services essential to human survival. Over the past 50 years, humans have transformed ecosystems more extensively and rapidly than at any other time in history, primarily to meet growing demands for food, freshwater, timber, fiber, and fuel. This has resulted in substantial and largely irreversible biodiversity loss (Ws.10, Living Planet Index; Ws.11, IPBES Conceptual Framework), leading to the identification of a new epoch: the Anthropocene. Nevertheless, because ecosystem services are not typically traded in markets and are rarely expressed in units comparable to economic or industrial outputs, they are often neglected in policy decisions.

The European Union places significant emphasis on *nature-based solutions* (NbS), defined as solutions inspired by or modeled on nature, which maximize the provision of ecosystem services while simultaneously delivering environmental, social, and economic benefits and enhancing ecosystem resilience (i.e., the capacity to resist and recover from disturbance or environmental stress). NbS represent cost-effective, long-term strategies for restoring ecosystems degraded by anthropogenic activities.

These solutions may be implemented locally or at the landscape scale. Local approaches aim to restore ecosystem **functionality** and local service provision, while landscape-scale strategies emphasize **connectivity**, for example, through ecological corridors, which play a key role in rehabilitating degraded landscapes, improving service provision, and enhancing progress toward the SDGs. NbS can improve multiple ecosystem services: for water supply, by reducing surface runoff and enhancing infiltration and storage; for flood mitigation, by promoting infiltration processes; for soil water retention, by minimizing obstructions in drainage systems caused by plant material; for soil protection, by reducing erosion processes; and for water quality, by lowering suspended sediments and pollutant loads.

Green infrastructure forms a core component of NbS, comprising strategically planned networks of natural and semi-natural areas designed and managed to deliver a wide range of ecosystem services while safeguarding biodiversity in both urban and rural areas. Some examples of artificial structures that can enhance ecosystem service provision or support species vitality are ecological corridors, high-biodiversity areas serving as reproduction centers for plant and animal species, and transitional zones that increase habitat availability, ecological quality, and biodiversity.

5.3 ECOSYSTEM PROCESSES, CLIMATE CHANGE, AND ECOLOGICAL TRANSITION

5.3.1 Photosynthesis and the carbon cycle

Most nutrients originate either from rocks in the Earth's crust or from the atmosphere. During ecosystem functioning, nutrients are repeatedly used by plants, animals, fungi, and microorganisms before being partly dispersed into watercourses and aquifers or returned to the atmosphere in gaseous form.

Organisms utilize chemical compounds to build their structures through specific metabolic processes. A general model of the **biogeochemical cycle** of an element in a forest ecosystem shows that plant and animal organic matter (leaves, detritus, and dead bodies) is recycled through decomposition processes. Nutrients released into the soil through mineralization are reabsorbed by plant roots, contributing to the formation of new plant tissues. This cycling of elements within the ecosystem ensures the continuous production of organic matter through **photosynthesis**—that is, the fixation of carbon—and through the conversion of inorganic nitrogen to organic forms.

In addition to nutrient cycling within ecosystems, nutrient losses must also be considered. These include the removal of organic detritus by rainfall and soil erosion, processes that vary according to the geomorphological characteristics of the landscape (e.g., the presence of slopes more or less steep). Leaching, or the migration into solution of minerals resulting from the decomposition of organic matter and their release into the soil, may occur through surface runoff or infiltration into the soil profile. Such nutrient losses (*outputs*) are balanced, at least in part, by nutrient inputs (*inputs*), including the weathering of minerals and the geological alteration of bedrock.

One of the most fundamental biogeochemical cycles underlying the functioning of terrestrial and aquatic ecosystems is the **carbon cycle**. This cycle is influenced by climate and ecological characteristics of the site, and it can be described across multiple levels of biological organization, from the molecular scale to ecosystems and landscapes, as well as globally at the biosphere level. In the atmosphere, carbon occurs as carbon dioxide (CO_2), a molecule assimilated by photosynthesis and released by respiration in both terrestrial and marine ecosystems (the latter mainly through phytoplankton and algae). The reaction of photosynthesis is:

$$6\,CO_2 + 6\,H_2O + \text{light energy} \rightarrow C_6H_{12}O_6 + 6O_2$$

$$N\ molecules\ of: 6\ \text{carbon dioxide} + 6\ \text{water} + (\text{light energy}) \rightarrow 1\ \text{glucose} + 6\ \text{oxygen}$$

In all aerobic organisms, cellular respiration produces CO_2 as a metabolic by-product.

These two processes—photosynthesis and respiration—were historically in balance, but over recent decades this equilibrium has been altered by anthropogenic CO_2 emissions, primarily from fossil fuel combustion for energy production.

The carbon cycle thus describes the movement of carbon in its various forms among the biosphere, atmosphere, and geosphere. Carbon reservoirs exchange carbon through a range of processes, that is, terrestrial plants absorb atmospheric CO_2 during photosynthesis and release it during respiration via stomatal gas exchange (stomata are generally located on the abaxial surface of leaves); an exchange process of CO_2 also occurs between the atmosphere and the oceans, where phytoplankton and algae assimilate dissolved CO_2 during photosynthesis.

Two additional processes, both of anthropogenic origin, play a critical role in altering the carbon cycle: fossil fuel combustion and land-use change. Fossil coal, natural gas, petroleum, and derivatives are consumed in industrial production, transportation, and power generation, all of which release CO_2 during combustion. Land-use change, a broad term, encompasses numerous human activities such as agriculture, deforestation, afforestation, and urbanization. As a consequence of these processes, atmospheric CO_2 concentrations have risen dramatically—from approximately 270 parts per million (ppm) in the pre-industrial era to approximately 429 ppm in February 2026, as recorded by atmospheric monitoring stations (Ws.12, Daily CO_2).

5.3.2 Climate change

The greenhouse effect is a natural process related to the Earth's energy balance, namely the ratio between incoming flux (incident solar radiation) and outgoing flux (reflected solar and infrared radiation emitted by the Earth's surface). The energy balance contributes to maintaining climatic conditions favorable to humans and to other living species. Greenhouse gases, due to their increasing atmospheric concentrations, intensify the natural greenhouse effect, leading to greater absorption of outgoing infrared radiation and, consequently, the warming of the lower atmosphere.

The most common greenhouse gases include water vapor, carbon dioxide, ozone, methane, nitrous oxides, and chlorofluorocarbons. Anthropogenic emissions primarily concern CO_2, released during the combustion of fossil fuels and from industrial activities such as cement production, as well as methane (CH_4), which originates from natural sources, including wetlands and rice fields, and from digestive processes in animals. Anthropogenic methane emissions are associated with oil and natural gas extraction, pipeline leakages, forest fires, anaerobic waste decomposition, intensive livestock farming, and coal mining.

The increase in atmospheric CO_2 and other greenhouse gases influences atmospheric warming, which is also amplified by water vapor: higher temperatures enhance evaporation rates, leading to greater humidity, thereby reinforcing global warming. Elevated atmospheric CO_2 concentrations not only raise the temperature of large water bodies but also alter their chemical-physical properties by increasing acidity through the formation of carbonic acid (H_2CO_3). The oceans absorb about 30% of annual anthropogenic CO_2 emissions, resulting in significant ocean acidification.

Land-use and land-cover changes also affect the carbon cycle: areas covered by forests or grasslands have markedly different impacts compared to impervious surfaces created by anthropogenic activities, such as industrial, urban, or commercial areas. Land-use change exerts profound influence not only on the carbon cycle but also on other biogeochemical cycles, such as those of water and nitrogen, and on the provision of ecosystem services. Another major source of CO_2 emissions is wildfires: more than 80% of wildfire-related emissions originate in tropical regions and in the Global South, unlike fossil fuel-related emissions.

In Italy, total greenhouse gas emissions (expressed as CO_2 equivalents) in 2023 decreased by 26.4% compared to the 1990 baseline, in line with Kyoto Protocol commitments. This decline, particularly evident since 2008, is attributable to reduced energy consumption and industrial output due to the economic crisis, the relocation of certain production sectors, the expansion of renewable energy sources, and improvements in energy efficiency. Compared to 1990, the energy production and transportation sectors remain the most impactful: transport-related emissions show a slight increase (about 2%), whereas emissions from energy production and industrial facilities have declined (about 30%).

The principal changes observed because of the greenhouse effect concern air and ocean temperatures across nearly the entire globe, which are now significantly higher than they were a century ago. Global temperatures reached record highs between 2015 and 2024, making it the warmest decade on record. In 2024, global average temperatures temporarily exceeded the critical threshold of 1.5°C above pre-industrial levels (1850–1900), the warmest year on record to date. Moreover, over the last 50 years, numerous extreme weather and climate events have intensified: some regions now experience more frequent heatwaves and heavy precipitation events, while regional patterns exhibit significant variability.

With few exceptions, glaciers are retreating globally, as are Greenland's ice sheets over the past two decades. The Arctic has warmed substantially in the last 50 years, and it is virtually certain that Arctic sea ice extent declined by approximately 12.5% per decade over the period 1979–2021. The particularly rapid summer retreat of sea ice (a 9.4–13.6% decrease per decade) is likely unprecedented in the past 1,500 years. Snow cover in the Northern Hemisphere has diminished, especially in spring, and permafrost has undergone thawing across many regions; conversely, after a slight decadal increase between 1979 and 2014, Antarctic sea ice has begun to decline dramatically.

Scenarios described by the Intergovernmental Panel on Climate Change (IPCC) in the Fifth Assessment Report (2018) present four possible CO_2 concentration pathways, referred to as representative concentration pathways (RCPs), derived from combinations of integrated assessment models, atmospheric chemistry models, and global carbon cycle models.

The IPCC's latest report, the Sixth Assessment Report (AR6), was completed between 2021 and 2023 (Ws.13, IPCC).

Across all SSP scenarios (from SSP1–1.9 to SSP5–8.5), atmospheric surface warming is projected between +0.89 and +6.4°C relative to 1850–1900, with associated uncertainty ranges (5th–95th percentile confidence intervals). Warming is expected to be greater over land than oceans, with the Arctic warming more than the global average.

Given these are global mean temperatures, it is virtually certain that by the end of the 21st century, most regions will experience more exceptionally hot days and fewer unusually cold days. The likelihood of longer, more frequent heatwaves exceeds 90%, although occasional, unusually cold winters may still occur. Dry regions are projected to become drier, while humid regions will become wetter. The probability that extreme precipitation events will increase in frequency and intensity in mid-latitudes and tropical wet regions exceeds 90%. Monsoon-affected areas are also projected to expand, with more intense rainfall and longer monsoon seasons.

Since biodiversity underpins ecosystem functional processes (Ecosystem services, biodiversity, and human well-being section), it is relevant to note the principal global drivers of biodiversity change: land-use change, climate change, nitrogen deposition, introduction of alien species, and rising atmospheric CO_2 concentrations. In the Mediterranean Basin, biodiversity loss is primarily attributable to land-use change and invasive species, followed by climate change, nitrogen deposition, and increasing atmospheric CO_2 (Sala et al., 2000).

Within this framework, remote sensing analyses have been employed to assess drought-induced stress phenomena, including forest dieback. Allen et al. (2010) illustrate the relationships between drought duration (short/prolonged period), drought intensity (low/high), temperature (cold/warm), and precipitation (humid/arid). Notably, while a relatively small tree mortality is observed with the current climate, projections for future climate indicate an increase in extreme drought conditions associated with climate change. This suggests a potential rise in the risk of global forest mortality driven by water and heat stress.

5.3.3 Ecological transition

According to a survey conducted across the European Union since 2017, 92% of European citizens consider climate change to be a serious issue. Recently, the IPCC will be at the thirtieth Conference of the Parties (COP30) of the United Nations Framework Convention on Climate Change (Ws. 14, IPCC). The IPCC is currently in its seventh assessment cycle, which formally began in July 2023 (Ws. 15, IPCC). The following link shows an infographic summary report, Climate Change 2023 (AR6), (Ws. 16, IPCC Explainer).

Such trends will have severe consequences for both European and global economies, infrastructure, food production, public health, biodiversity, and political stability. For example, by 2100, the annual economic damage caused by river floods in Europe is projected to rise from €5 billion to €112 billion, while 16% of areas currently characterized by a Mediterranean climate could experience heightened aridity.

The European Commission's strategic vision responds to the Paris Agreement's call to continue efforts to limit global warming to 1.5 °C above pre-industrial levels, and it is fully aligned with the United Nations' SDGs. Specifically, achieving a greenhouse gas emission-free economy requires seven core strategic elements (Going Climate Neutral by 2019):

- Maximizing the benefits of energy efficiency, including zero-emission buildings;
- Optimizing the deployment of renewable energy sources and electrification to fully decarbonize Europe's energy supply;
- Adopting clean, safe, and connected mobility systems;
- Recognizing the competitiveness of European industry and the circular economy as key factors in reducing greenhouse gas emissions;
- Developing smart grid infrastructure and adequate interconnections;
- Fully exploiting the benefits of the bioeconomy and creating the necessary carbon sinks;
- Addressing residual CO_2 emissions through carbon capture and storage (CCS) technologies.

These seven strategic elements will be implemented through major development pathways, such as:

- Stimulating research and innovation;
- Increasing private investment;

- Providing clear signals to markets;
- Ensuring social cohesion so that no one is left behind.

The transition to a net-zero greenhouse gas economy is not only a matter of technology and employment but also of people, lifestyles, mobility choices, and modes of living and working together. Individual decisions influence both carbon and ecological footprints—from purchasing a car to dietary habits to housing choices. Lifestyle changes can facilitate the transition to climate neutrality while simultaneously enhancing quality of life.

Cities are laboratories for sustainable and transformative solutions. Urban regeneration and planning can also drive housing renovation, encouraging people to relocate closer to workplaces, thereby improving living conditions and reducing travel times and associated stress.

The European Union is committed to mobilizing the required investments, with 20% of the current budget already allocated to climate-related actions. The European Green Deal seeks to transform the EU into a modern, resource-efficient, and competitive economy (Ws.17, Green Deal).

One-third of the €1.8 trillion investment package under the Next Generation EU recovery plan, along with the EU's seven-year budget, will finance initiatives aimed at achieving this goal.

Within the framework of Horizon Europe, the EU's research and innovation program for 2021–2027, the Commission has proposed allocating 35% of its €100 billion budget to climate-related objectives through the development of low-cost innovative solutions.

The commission's proposal for a climate-neutral transformation encompasses sectors such as:

- Electrification (e.g., renewable energy, smart grids, and batteries).
- Hydrogen and fuel cells.
- Energy storage systems.
- Zero-carbon transformation of energy-intensive industries.
- Circular economy.
- Bioeconomy.
- Sustainable agricultural and forestry practices.

All these sectors will be supported by emerging technologies, including advanced materials, digitalization, artificial intelligence, information systems, and biotechnologies. Overall, achieving a climate-neutral economy is not expected to negatively impact economic prospects. On the contrary, projections suggest that by 2050—once fully decarbonized—the EU economy will have more than doubled in size compared to 1990 (Ws.18, Green transition).

5.4 ECOSYSTEM SERVICES: FROM NATURE-BASED SOLUTIONS TO THE NATURE-BASED ECONOMY

5.4.1 The socio-ecological system

In relation to the 17 SDGs of the United Nations 2030 Agenda, it is recalled that economy and ecology are not always aligned: economic choices frequently diverge from environmental protection policies. However, to achieve sustainable development and to improve or maintain human well-being, it is necessary to integrate socioeconomic and ecological systems into a unified socio-ecological system, thereby attaining harmony between society and nature (Grove, 2009). Sustainable development is defined as the capacity to meet present needs without compromising the ability of future generations to meet their own (Brundtland, 1987).

A central concept in this framework is natural capital, a term borrowed from the economic sector, which refers to the stock of natural resources—including plants, animals, air, water, soil, and minerals—generated through the flows of matter and energy within ecosystems that collectively provide benefits to humans (Ws.19, IPBES Conceptual Framework). It is thus crucial to examine the interaction between natural capital and human, socio-cultural, and infrastructural capital as a determinant of human well-being: infrastructural and human capital (the economy) are embedded within society, which is in turn embedded within nature. The contribution of natural capital to human well-being is constituted by ecosystem services, and the analysis and management of natural resources require a transdisciplinary approach. Biodiversity remains central: the variety of genes, species, and functional traits in ecosystems plays a fundamental role in ecosystem functioning and, in the goods and services that ecosystems provide to humanity.

Considering current planetary boundaries (Ws.20, GLOBAIA), climate change is particularly illustrative. The month of December 2025 was the fifth warmest on record globally as reported by *Copernicus Climate Change Service* (Ws.21, Copernicus).

As discussed in Section 5.3.2, this anomaly is primarily attributable to increasing concentrations of greenhouse gases such as carbon dioxide and methane, generated by anthropogenic sources including industrial activity, agriculture, waste disposal, intensive livestock farming, road traffic, energy production, and air and maritime transport, particularly from urban centers. Atmospheric CO_2 concentrations have steadily risen; for instance, emissions recorded in autumn–winter 2006 demonstrated high output, predominantly in the Northern Hemisphere (North America, Europe, and China). From 1990 (the baseline year established under the Kyoto Protocol) to 2019, CO_2 emissions from major industrialized countries did not decrease; on the other hand, a sharp rise occurred around 2002 in China and India.

Emissions gaps 2035 (Ws.22, Global emissions gaps), showing global annual greenhouse gas emissions (Gt CO_2 eq/year) from 1990 to 2025, with projections to 2035. According to this forecast (November 2025), compliance with the Paris Agreement (COP21, 2015)—which aims to limit warming to 1.5°C—would require a reduction of 26–31 Gt CO_2 eq by 2035.

The EU's current emissions reduction target is to achieve climate neutrality (net-zero CO_2 emissions) by 2050 across sectors such as agriculture, services, transport, industry, and energy production.

A second planetary boundary of concern is biodiversity, encompassing terrestrial and aquatic ecosystems, where species richness is under severe threat from anthropogenic activities. According to the IPBES 2019 report, the primary direct drivers of biodiversity loss across terrestrial, freshwater, and marine ecosystems include land-use change and overexploitation of natural resources, which together account for more than 50% of impacts. Climate change, pollution, and the spread of invasive alien species represent additional drivers. Approximately 47% of ecosystems exhibit declining trends in extent and health; around 25% of species are at risk of extinction, and ecological communities show reduced integrity; the abundance of naturally occurring species in ecosystems has declined by an average of 23% in terrestrial communities. Globally, biomass and abundance of species, particularly wild mammals, have declined by about 82%; vertebrate abundance indicators have dropped rapidly since 1970. Furthermore, natural resources vital to indigenous and local communities have deteriorated across nearly 72% of conservation status indicators.

In summary, biodiversity and ecosystem services essential to human survival are experiencing global degradation due to direct and indirect anthropogenic impacts, which have intensified over the past five decades. Current trajectories make it impossible to achieve conservation and sustainable use targets without significant transformations in socioeconomic, political, and technological choices. Natural resources must therefore be exploited sustainably while simultaneously addressing other societal challenges through concrete, coordinated, and immediate actions.

A third planetary boundary involves the quality of air, water, and soil matrices. Environmental pollution, particularly air pollution, undermines ecosystem health and human well-being. Pollutants include greenhouse gases and are classified as primary (directly released from anthropogenic sources) or secondary (formed through reactions among primary pollutants). The European Environment Agency (EEA) 2024 report on air quality estimates that 357,000 premature deaths in 2022 in the EU27 were attributable to pollutants such as $PM_{2.5}$, nitrogen dioxide, and ozone. For Italy, the most recent data indicate approximately 48,600 premature deaths from $PM_{2.5}$, 9,600 from nitrogen dioxide, and 13,600 from ozone. The EU Green Week 2021 was dedicated to the theme *Zero Pollution: Healthier People and Planet.*

These issues are particularly critical in urban socio-ecosystems, given projections of a 2.5 billion increase in the global urban population by 2050, with 404 million additional urban inhabitants in India, 292 million in China, and approximately 90 million in the United States (McPhearson et al., 2016). In Europe, around 75% of the population currently resides in urban areas, making urban socio-ecological systems crucial for analysis, management, and monitoring due to the multiple environmental pressures characteristic of metropolitan areas at global, regional, and local scales. At the local scale, urban ecosystems—through land-use and land-cover change (urbanization)—drive alterations in biogeochemical cycles, air pollution, urban heat island effects, and biodiversity loss, thereby exerting cascading impacts at broader scales.

In this context, the 2030 Agenda is particularly relevant, especially Goal 11 (Sustainable Cities and Communities), along with other targets advocating inclusive, sustainable urbanization and participatory, integrated planning by 2030. European strategies aligned with these objectives include the EU Biodiversity Strategy for 2030, the EU Forest Strategy, the Green Infrastructure Strategy, the Climate Adaptation Strategy, the recent Natural Restoration Law, and the Urban Agenda for Europe—adopted by Member States—as well as the Italian National Urban Green Strategy, which emphasizes the need for municipalities to adopt planning and projects based on ecosystem services while strengthening the functions of green infrastructure networks.

5.4.2 Metropolitan areas and sustainable management

Cities represent the predominant form of human settlement in the contemporary era, functioning both as pivotal centers of human activity and as key drivers of alteration of environmental balances (Grimm et al., 2008; Grove, 2009; Solecki et al., 2013). Urban socio-ecological systems are characterized by high population density, significant land-use change, and the use of natural resources that are not locally available. In Europe, urbanization is proceeding rapidly, causing soil sealing in peri-urban areas (EEA, 2006) and leading to a reduction in soil quality and functionality.

Urban expansion primarily occurs at the expense of suburban and rural areas, typically characterized by low population density (Brueckner, 2000). Consequently, agricultural land, pastures, and forests are increasingly absorbed or converted into extensive metropolitan areas. Over the last thirty years, the size of urban areas in Europe has increased by 5.4%, mostly in regions with high population density and intensive economic activity.

In relation to the expansion of urbanized territories, it is important to highlight that urban green spaces generate a wide range of benefits for citizens: cultural-historical values inherent to green areas, socio-health benefits, improvements in domestic and occupational environments with positive impacts on physical and mental health, recreational opportunities, aesthetic enrichment of the landscape through variation in plant color, shape, and density, as well as opportunities for direct contact with nature and seasonal dynamics.

Environmental benefits include mitigation of urban summer temperature (urban heat island effect), carbon sequestration, removal of atmospheric pollutants (PM_{10}, $PM_{2.5}$, and O_3), noise absorption, and provision of habitats for wildlife. By reducing atmospheric pollutants and mitigating local climatic conditions, urban vegetation provides tangible health benefits, lowering both hospitalization rates and societal costs (Manes et al., 2008; Mannocci et al., 2014; Orioli et al., 2019).

Economic benefits include the increase in property values and the promotion of tourism.

The Common International Classification of Ecosystem Services (Ws.23, CICES), developed by the EEA, has proposed a hierarchical classification system designed to facilitate environmental accounting for regulating services. This framework can also be applied to urban ecosystems, particularly with respect to carbon sequestration, air pollutant removal, and local climate regulation.

Air pollution is among the primary drivers of climate change and exerts significant negative impacts on human health. In Europe, between 1980 and 2010, damages from hurricanes and floods—extreme weather events associated with climate change—were estimated at approximately € 300 billion, with 75,000 heat-related deaths recorded during the same period. Globally, Vohra et al. (2021) estimated that in 2012 alone, approximately 10.2 million premature deaths could be attributed to fine particulate air pollution ($PM_{2.5}$). Atmospheric pollutants also contribute to ecosystem degradation, biodiversity loss, and the reduction of ecosystem services. For example, globally, 20% of drinking water in urban areas is polluted.

To mitigate the impacts of pollution and its associated economic costs, "NbS" have been increasingly adopted to foster sustainable development, enhance resilience, and promote green growth while supporting the creation of new employment opportunities.

Regarding soil–plant–atmosphere relationships and interactions with air pollutants, in the case of a plant—whether a herbaceous, shrubby, or arboreal species—as already stated, the main physiological processes include photosynthesis and transpiration within the canopy layer. These involve the assimilation of CO_2 and concurrent production of oxygen, as well as stomatal transpiration, through which plants emit water vapor. Urban vegetation (parks, gardens, street trees, etc.) plays a crucial role in the absorption of gaseous air pollutants such as ozone (a strong oxidizing agent) via leaf stomatal exchange, as well as in the adsorption of atmospheric particulates (PM_{10}, $PM_{2.5}$) on leaf surfaces (Fares et al., 2013, 2016; Fusaro et al., 2017, 2018). Urban greenery can therefore improve environmental quality through

Table 5.1 Estimates of ozone (O_3) and particulate matter (PM_{10}) removal by vegetation for ten Italian metropolitan cities

		Removal of O_3		Removal of PM_{10}	
	Forested area (ha)	*Normalized ($g \times m^{-2}$)*	*Total (Mg)*	*Normalized ($g \times m^{-2}$)*	*Total (Mg)*
Bari	17,814.9	2.6	462.9	0.38	68.9
Bologna	79,008.4	3.48	2,751.4	1.5	1,194.9
Florence	169,891.4	3.64	5,921.5	1.66	2,712.5
Genoa	120,945.7	4.68	5,657.9	2.0	2,435.2
Milan	5,481.5	2.66	146.1	1.54	84.8
Naples	17,077.3	3.37	575.6	2.94	504.0
Reggio Calabria	129,321.8	3.36	4,344.3	2.04	2,647.7
Rome	111,955.5	3.57	3,999.7	2.08	2,319.8
Turin	195,149.0	3.14	6,136.8	1.18	2,320.5
Venice	861.6	2.0	18.4	1.34	11.7
Total	847,507.1		30,014.6		14,299.9

Source: Manes et al. (2016), updated data.

pollutant removal, and these regulating ecosystem services can also be monetized by attributing values to avoided negative externalities.

A study conducted in ten Italian metropolitan cities (Manes et al., 2016) analyzed the capacity of urban and peri-urban forests to remove PM_{10} and ozone. Table 5.1 reports pollutant removal data for the ten cities.

Using satellite remote sensing data, it was possible to estimate the quantities of pollutants removed and to assign a monetary value to this regulating ecosystem service. Instead of using the market price of goods or services that caused pollution, the study relied on the estimated social costs of the pollutants' adverse effects on human health and the environment, expressed as the value per ton of pollutants removed (Roy et al., 2012). According to the EEA (2021), the estimated value per ton removed was approximately € 107,384 for PM_{10} and €17,110 for ozone. In summary, one hectare of urban forest removes on average 17 kg/year of PM_{10}, equivalent to €1,825, with an overall estimated economic value of €1.532 billion for the ten metropolitan cities in 2003. Similarly, one hectare of urban forest removes on average 35.7 kg/year of ozone, equivalent to € 611, for a total estimated value of € 512 million (2003). It is important to underline that, for example, the removal of 1 ton of PM_{10} in the recent EEA report (2023) is valued at €154,330.00.

In the Metropolitan City of Rome, several studies (Calfapietra et al., 2013; Cattani et al., 2017; Manes et al., 2012; Manes et al., 2016; Sebastiani et al., 2021) evaluated seasonal dynamics and species-specific pollutant removal capacity. Seasonal PM_{10} deposition calculated across different vegetation functional groups (deciduous broadleaves, evergreen broadleaves, and conifers) corresponded to a monetary value of € 221 million in 2015. These findings highlight the importance of synergistic interactions among different functional groups in Mediterranean climates, underscoring the necessity of biodiversity conservation to enhance the efficiency of ecosystem service provision.

Structural traits such as *LAI* (m^2 of leaves per m^2 of ground) and functional traits such as stomatal conductance (degree of stomatal aperture regulating gas exchange) represent the primary factors influencing the removal of particulate matter and ozone by urban and peri-urban forests. The synergistic contribution of different vegetation types is further reinforced by seasonal responses: deciduous species exhibit high pollutant removal during spring and summer, whereas evergreen species maintain higher annual capacity due to their ability to function during winter months. These findings are consistent with international studies.

Among regulating services, local climate mitigation is particularly important. As previously discussed, cities are characterized by the urban heat island effect, with temperatures up to 12°C higher than surrounding rural areas. Urban vegetation mitigates this phenomenon through shading and leaf transpiration.

A study conducted in Rome using Landsat eight satellite data combined with in-situ observations delineated a transect from the coastal dune system to the urban core, crossing peri-urban forests (Castelporziano Presidential Estate), agricultural fields, and historic green spaces such as Villa Ada, as well as a section of the Tiber River (Marando et al., 2019). During summer, Land Surface Temperature (LST) was highest in unvegetated sandy dunes; it decreased significantly in the peri-urban forest (Castelporziano); it rose again in fallow lands and cultivated fields, as in the summer period these are generally areas devoid of vegetation with bare soil; and, as expected, it peaked in densely urbanized areas and declined once more within green infrastructures such as Villa Ada's forest and along the Tiber River (a blue infrastructure), where reductions of 7–8 °C were recorded. These findings demonstrate that vegetated areas play a crucial role in regulating local summer climate: average LST values in urban and peri-urban forests in Rome were approximately 7°C lower than in urban fabric. Vegetation indices such as NDVI, derived from satellite imagery, provide effective indicators of local climate regulation.

The dynamics of metropolitan land-use change should be at the forefront of national and global governance. According to the IPCC, urban areas collectively contribute approximately 70% of CO_2 equivalent emissions from final energy consumption, a share expected to rise to 76% by 2030.

Management efforts should embrace a *One Health* perspective, enhancing ecological connectivity within peri-urban belts and between urban green spaces and surrounding natural areas, while reducing soil consumption. The *One Health* framework, adopted by WHO (human health), FAO (sustainable food and agriculture), WOAH (animal health), and UNEP (environmental protection for future generations), emphasizes an integrated and unifying approach that considers the interdependence of human, animal, plant, environmental, and ecosystem health. This conceptual and operational approach requires multisectoral, interdisciplinary, and cross-societal collaboration to address complex challenges.

It is necessary to implement ecological restoration and environmental rehabilitation initiatives, as mandated by the **Nature Restoration Regulation**, definitively approved by the Council of Europe, and enacted on August 18, 2024. This regulation aims to enhance ecosystem quality across Europe, where approximately 80% of terrestrial habitats are in various states of degradation, thereby strengthening the provision and multifunctionality of regulating, provisioning, and cultural ecosystem services.

Finally, ecological restoration of terrestrial, coastal marine, and freshwater ecosystems requires transdisciplinary expertise to integrate social, economic, and environmental dimensions into a comprehensive approach.

5.4.3 Green infrastructure and nature-based solutions

As highlighted in both the Italian National Urban Green Strategy and the EU Biodiversity Strategy for 2030, enhancing the provision of ecosystem services requires increasing the surface area covered by urban and peri-urban forests. Reforestation interventions should be based on the principle of expanding urban and peri-urban forests "from square meters to hectares."

In this regard, it is also important to introduce the concept of *ecosystem service **supply areas*** and *service **demand areas***. Data obtained for the Metropolitan City of Rome reveal zones with high levels of urbanization and others characterized by extensive peri-urban forests—such as the Presidential Estate of Castelporziano—or by urban forests, such as historical villas. We can thus identify *supply areas*, that is, areas providing ecosystem services, and *demand areas*, that is, areas where ecosystem services are needed, typically where green spaces are absent or limited in size, and population density is high. Consequently, residents in areas with high demand and low supply are exposed, for example, to higher levels of air pollution, with negative impacts on health and quality of life.

It is possible to estimate this imbalance (*mismatch*) between the demand and supply of ecosystem services in a given territory. A study conducted for the Municipality of Rome (Sebastiani et al., 2021) identified this mismatch by integrating multidisciplinary data and knowledge to support sustainable urban development, contributing to the definition of urban forestry plans promoted under the Next Generation EU initiative. This study enabled the identification of priority intervention areas where the mismatch is most pronounced—mainly residential or industrial zones with a limited presence of open and green spaces. The compact urbanization model, long encouraged for its reduced impact on land consumption, may in fact negatively affect the balance between the supply and demand of ecosystem services.

It is therefore advisable to implement urban regeneration and reforestation projects, both in existing green areas and in abandoned zones—especially where soil de-sealing is not feasible—and to promote initiatives aimed at protecting nature, restoring degraded ecosystems, and increasing urban and peri-urban forests through the enhancement of green infrastructure networks for the sustainable development of metropolitan areas, as recently indicated in the Nature Restoration Law for the countries of the European Community, which calls for the planting of 3 billion trees in Europe by 2030.

Each topic discussed in this chapter has recalled several of the SDGs of the 2030 Agenda, particularly SDGs 1, 3, 4, 7, 11, 13, 14, and 15. Regarding SDG 11, international research has focused on specific targets to be achieved by 2030 in order to fully realize this goal. The main target is to

> Reduce the per capita environmental impact of cities, paying particular attention to air quality and municipal and other waste management; provide universal access to safe, inclusive, and accessible green and public spaces, especially for women, children, older persons, and persons with disabilities; and substantially increase the number of cities and human settlements adopting and implementing integrated policies and plans aimed at inclusion, resource efficiency, climate change mitigation, and adaptation.

This discussion helps to understand how the ecological transition is currently unfolding in Europe. In 2013, within the framework of the EU Biodiversity Strategy, the MAES project (*Mapping and Assessment of Ecosystems and their Services*; Maes et al., 2012) was launched, emphasizing the importance for European countries to map ecosystem services and assess their supply and demand (Ws.24, MAES).

In line with the objectives set by the European strategy, Italy adopted a National Biodiversity Strategy structured around three main themes: biodiversity and ecosystem services, biodiversity and climate change, and biodiversity and economic policies. In particular, the Italian Natural Capital Committee was established, tasked with gathering results on the assessment of natural capital and ecosystem services at the national scale, analyzing the pressures threatening their conservation status, and suggesting actions to mitigate or prevent their loss. Equally important is the role of the *Public Green Committee*, which has recently produced specific guidelines.

It is essential to protect and restore natural ecosystems and increase urban and peri-urban forests through the enhancement of green infrastructure networks—key activities for the sustainable development of metropolitan areas. This is underscored also by the EU Biodiversity Strategy for 2030, which calls for the planting of 3 billion trees across Europe, as already specified by NRL. The Italian *National Recovery and Resilience Plan (PNRR)* foresees the planting of 6.6 million trees in the 14 Italian metropolitan cities.

Furthermore, it is crucial to maintain the role of both structural and functional biodiversity and to prioritize the use of native species, selected according to the biogeographical characteristics of the areas involved in reforestation, taking into account specific ecosystem service needs. It is also desirable to define transdisciplinary guidelines to support decision-makers and stakeholders in implementing specific nature-based solutions (Ws.25, NbS), (Maes & Jacobs, 2017; Marando et al., 2016), aimed at improving human health and well-being in urban areas, as demonstrated by the results of a recent international project conducted in many European cities (Ws. 26, EnRoute; *EnRoute – Final Conference, Brussels, 29 February 2019*).

Finally, it is important to remind that, according to the *Do No Significant Harm (DNSH)* principle, achieving sustainable development requires harmonizing three fundamental dimensions: environmental protection, economic growth, and social inclusion. Human prosperity and well-being depend closely on the sustainable use of natural capital, which ensures the generation of multiple ecosystem services essential for human health. The amendment to Articles 9 and 41 of the Italian Constitution, which now includes the protection of the environment, biodiversity, and ecosystems among the fundamental principles of the Republic, clearly reflects a commitment to the principle of integral ecology.

To preserve resources for future generations, in accordance with the UN Pact for the Future, a deep understanding of ecological processes is required, as well as the proper economic valuation of natural capital and ecosystem services. Moreover, it is essential to promote the intrinsic value of biodiversity—which cannot be translated into monetary terms—reflecting its cultural and ethical significance. Respecting the DNSH principle requires territorial interventions that do not directly or indirectly cause biodiversity loss or alter ecosystem functioning, thereby reducing or eliminating the goods and services ecosystems provide.

SUGGESTED TEACHING TEXTS

Cain, M. L., Bowman, W. D., & Hacker, S. D., 2017. *Ecologia*, PICCIN.

Odum, E. P., 1963. Primary and secondary energy flow in relation to ecosystem structure. Washington DC, *Proceedings of the 16th International Congress of Zoology*, pp. 336–338.

Odum, E. P., & Barret, G. W., 2007. *Fondamenti di Ecologia*, PICCIN.
Sadava, D., Hillis, D. M., Heller, H. C., & Hacker, S., 2019. *Biologia Volume 6 – L'ecologia*, Zanichelli Ed.
Smith, R. L., & Smith, T. M., 2017. *Elementi di Ecologia*, PEARSON.

SUGGESTED SITES

Ws.1 - IPBES, https://www.ipbes.net/conceptual-framework
Ws.2 – IPCC, https://www.ipcc.ch/report/ar5/wg1/carbon-and-other-biogeochemical-cycles/
Ws.3 - EU Soil Strategy for 2030, https://environment.ec.europa.eu/topics/soil-health/soil-strategy-2030_en
Ws.4 – MEA, https://www.millenniumassessment.org/en/index.html
Ws.5 - UN Agenda 2030, https://www.un.org/en/conferences/SDGSummit2023/about/background
Ws. 6 – GLOBAIA, https://globaia.org/boundaries
Ws.7 - IPBES Secretariat, https://files.ipbes.net/ipbes-web-prod-public-files/downloads/Decision%20IPBES_2_4.pdf
Ws.8 - MEA, https://www.millenniumassessment.org/en/Condition.html
Ws.9 – TEEB, https://www.unep.org/topics/teeb
Ws.10 - Living Planet Index, https://www.livingplanetindex.org/
Ws.11 – IPBES Conceptual Framework, https://www.ipbes.net/conceptual-framework
Ws.12 – Daily CO2, https://co2.earth/daily-co2
Ws.13 – IPCC, https://www.ipcc.ch/report/sixth-assessment-report-cycle/
Ws.14 – IPCC, https://www.ipcc.ch/event/ipcc-at-cop30/
Ws.15 – IPCC, https://www.ipcc.ch/assessment-report/ar7/
Ws.16 - IPCC Explainer, https://eciu.net/analysis/infographics/ipcc-explainer-infographic-synthesis-report-climate-2023
Ws.17 – Green Deal, https://commission.europa.eu/topics/climate-action/delivering-european-green-deal_en
Ws.18 – Green transition, https://reforms-investments.ec.europa.eu/technical-support-instrument-0/green-transition_en
Ws.19 – IPBES Conceptual Framework, https://www.ipbes.net/conceptual-framework
Ws.20 – GLOBAIA, https://globaia.org/boundaries
Ws.21 – Copernicus, https://climate.copernicus.eu/
Ws.22 – Global emissions gaps, https://climateactiontracker.org/global/emissions-gaps/
Ws.23 - CICES, https://cices.eu/revision-highlights/
Ws.24 – MAES, https://biodiversity.europa.eu/europes-biodiversity/ecosystems/maes
Ws.25 – NbS, https://rea.ec.europa.eu/funding-and-grants/horizon-europe-cluster-6-food-bioeconomy-natural-resources-agriculture-and-environment/nature-based-solutions_en

Ws.26 – EnRoute, https://oppla.eu/case-study/greening-rome-human-and-ecosystem-health

REFERENCES

Allen, C. D., Macalady, A. K., Chenchouni, H., Bachelet, D., McDowell, N., Vennetier, M.,... & Cobb, N., 2010. A global overview of drought and heat-induced tree mortality reveals emerging climate change risks for forests. *Forest Ecology and Management*, 259(4), 660–684.

Braat, L. C., & De Groot, R., 2012. The ecosystem services agenda: Bridging the worlds of natural science and economics, conservation and development, and public and private policy. *Ecosystem Services*, 1(1), 4–15.

Brueckner, J. K., 2000. Urban Sprawl: Diagnosis and remedies. *International Regional Science Review*, 23, 160–171.

Brundtland, G., 1987. *Rapporto ONU della Commissione Mondiale su Ambiente e Sviluppo*. published by the UN World Commission on Environment and Development.

Calfapietra, C., Fares, S., Manes, F., Morani, A., Sgrigna, G., & Loreto, F., 2013. Role of Biogenic Volatile Organic Compounds (BVOC) emitted by urban trees on ozone concentration in cities: A review. *Environmental Pollution*, 183, 71–80.

Cattani, G., Gaeta, A., Di Menno Di Bucchianico, A., De Santis, A., Gaddi, R., Cusano, M., Ancona, A., Badaloni, C., Forastiere, F., Gariazzo, C., Sozzi, R., Inglessis, M., Silibello, C., Salvatori, E., Manes, F., Cesaroni, G., on behalf of the VIIAS study group, 2017. Development of land-use regression models for exposure assessment to ultrafine particles in Rome, Italy. *Atmospheric Environment*, 156, 52–60.

COP21, UN Climate Change Conference of Parties in Paris, France, 12 December 2015 https://unfccc.int/process-and-meetings/the-paris-agreement; UNFCCC, The Paris Agreement https://treaties.un.org/pages/ViewDetails.aspx?src=TREATY&mtdsg_no=XXVII-7-d&chapter=27&clang=_en

Costanza, R., De Groot, R., Sutton, P., Van der Ploeg, S., Anderson, S. J., Kubiszewski, I.,... & Turner, R. K., 2014. Changes in the global value of ecosystem services. *Global Environmental Change*, 26, 152–158.

Daily, G. C., 1997. Introduction: what are ecosystem services. *Nature's services: Societal Dependence on Natural Ecosystems*, 1(1), 1–10.

European Environment Agency, 2006 *EEA, Average annual change in soil sealing, 2006-2015*. https://www.eea.europa.eu/en/analysis/maps-and-charts/average-annual-increase-of-sealing-2; Soil sealing index Published: 26 Mar 2026, Temporal coverage: 2006-2021(https://www.eea.europa.eu/en/datahub/datahubitem-view/33029ced-43e7-421a-8df1-1660fd590b74)

European Environment Agency, 2023 EEA, Technical note_Estimating the external costs of industrial air pollution-Trends 2012-2021v2.pdf, https://www.eea.europa.eu/en/analysis/publications/the-costs-to-health-and-the-environment-from-industrial-air-pollution-in-europe-2024-update/technical-note_estimating-the-external-costs-of-industrial-air-pollution-trends-2012-2021v2.pdf

External costs of energy-intensive facilities, 2021, https://www.eea.europa.eu/en/analysis/publications/zero-pollution-decarbonisation-and-circular-economy-in-energy-intensive-industries/external-costs-of-energy-intensive-facilities-2021

Fares, S., Matteucci, G., Mugnozza, G. S., Morani, A., Calfapietra, C., Salvatori, E., Fusaro, L., Manes, F., & Loreto, F., 2013. Testing of models of stomatal ozone fluxes with field measurements in a mixed Mediterranean forest. *Atmospheric Environment*, 67, 242–251.

Fares, S., Savi, F., Fusaro, L., Conte, A., Salvatori, E., Aromolo, R., & Manes, F., 2016. Particle deposition in a peri-urban Mediterranean forest. *Environmental Pollution*, 218, 1278–1286.

Fusaro, L., Marando, F., Sebastiani, A., Capotorti, G., Blasi, C., Copiz, R., Congedo, L., Munafò, M., Ciancarella, L., & Manes, F., 2017. Mapping and assessment of PM_{10} and O_3 removal by woody vegetation at urban and regional level. *Remote Sensing*, 9(8), 791.

Fusaro, L., Mereu, S., Salvatori, E., Agliari, E., Fares, S., & Manes, F., 2018. Modeling ozone uptake by urban and peri-urban forest: A case study in the Metropolitan City of Rome. *Environmental Science and Pollution Research*, 25(9), 8190–8205.

Grimm, N. B., Faeth, S. H., Golubiewski, N. E., Redman, C. L., Wu, J., Bai, X., & Briggs, J. M., 2008. Global change and the ecology of cities. *Science*, 319, 756–760.

Grove, J. M., 2009. Cities: Managing densely settled social-ecological systems. In: Chapin, F. Stuart III, Kofinas, Gary P., & Folke, Carl, eds. *Principles of Ecosystem Stewardship. Resilience-Based Natural Resource Management in a Changing World*. Springer: 281–294.

Haines-Young, R., & Potschin, M., 2018. *Common International Classification of Ecosystem Services (CICES) V5.1*. Guidance on the Application of the Revised Structure.

Maes, J., Egoh, B., Willemen, L., Liquete, C., Vihervaara, P., Schägner, J.P., Grizzetti, B., Drakou, E. G., Notte, A. L., Zulian, G., Bouraoui, F., Luisa Paracchini, M., Braat, L., & Bidoglio, G., 2012. Mapping ecosystem services for policy support and decision making in the European Union. *Ecosystem Services*, 1, 31–39.

Maes, J., & Jacobs, S., 2017. Nature-based solutions for Europe's sustainable development. *Conservation Letters*, 10, 121–124.

Manes, F., Incerti, G., Salvatori, E., Vitale, M., Ricotta, C., & Costanza, R., 2012. Urban ecosystem services: Tree diversity and stability of tropospheric ozone removal. *Ecological Applications*, 22(1), 349–360.

Manes, F., Marando, F., Capotorti, G., Blasi, C., Salvatori, E., Fusaro, L., Ciancarella, L., Mircea, M., Marchetti, M., Chirici, G., & Munafò, M., 2016. Regulating ecosystem services of forests in ten Italian Metropolitan Cities: Air quality improvement by PM_{10} and O_3 removal. *Ecological Indicators*, 67, 425–440.

Manes, F., Salvatori, E., La Torre, G., Villari, P., Vitale, M., Biscontini, D., & Incerti, G., 2008. Urban green and its relation with air pollution: Ecological studies in the Metropolitan area of Rome. *Italian Journal of Public Health*, 5(4), 278–283.

Mannocci, A., Salvatori, E., Saulle, R., Incerti, G., Boccia, A., Manes, F., & La Torre, G., 2014. Health risks associated with urban environmental pollution in Europe: Results of the HEREPLUS multicentre study in Rome (Italy). *Igiene e Sanità Pubblica*, 70(5), 463–472.

Marando, F., Salvatori, E., Fusaro, L., & Manes, F., 2016. Removal of PM_{10} by forests as a nature-based solution for air quality improvement in the Metropolitan city of Rome. *Forests*, 7(7), 150.

Marando, F., Salvatori, E., Sebastiani, A., Fusaro, L., & Manes, F., 2019. Regulating Ecosystem Services and Green Infrastructure: assessment of Urban Heat Island effect mitigation in the municipality of Rome, Italy. *Ecological Modelling*, 392, 92–102.

McPhearson, T., Pickett, S. T., Grimm, N. B., Niemelä, J., Alberti, M., Elmqvist, T.,... & Qureshi, S., 2016. Advancing urban ecology toward a science of cities. *Bio Science*, *66*(3), 198–212.

Millennium Ecosystem Assessment, 2005. *Ecosystems and Human Well-Being: Biodiversity Synthesis*. World Resources Institute.

Orioli, R., Antonucci, C., Scortichini, M., Cerza, F., Marando, F., Ancona, C., Manes, F., Davoli, M., Michelozzi, P., Forastiere, F., & Cesaroni, G., 2019. Exposure to residential greenness as a predictor of cause-specific mortality and stroke incidence in the Rome longitudinal study. *Environmental Health Perspectives*, 127(2), 027002.

Roy, S., Byrne, J., & Pickering, C., 2012. A systematic quantitative review of urban tree benefits, costs, and assessment methods across cities in different climatic zones. *Urban Forestry and Urban Greening*, 11, 351–363.

Sala, O. E., Chapin, F. S., Armesto, J. J., Berlow, E., Bloomfield, J., Dirzo, R.,... & Wall, D. H., 2000. Global biodiversity scenarios for the year 2100. *Science*, 287(5459), 1770–1774.

Sebastiani, A., Marando, F., & Manes, F., 2021. Mismatch of regulating ecosystem services for sustainable urban planning: PM10 removal and urban heat island effect mitigation in the municipality of Rome (Italy). *Urban Forestry & Urban Greening*, 57, 126938.

Solecki W., Seto K. C., & Marcotullio P. J., 2013. It's time for an urbanization science, *Environment: Science and Policy for Sustainable Development*, 55(1), 12–17.

Vohra, K., Vodonos, A., Schwartz, J., Marais, E. A., Sulprizio, M. P., & Mickley, L. J., 2021. Global mortality from outdoor fine particle pollution generated by fossil fuel combustion: Results from GEOS-Chem. *Environmental Research*, 195, 110754.

Chapter 6

The law of sustainable development in the policies and regulations of the European Union: Relevant SDGs: 16, 7, 11, 12, 13, 14, and 15

Angelo Lalli

6.1 THE LAW FOR SUSTAINABLE DEVELOPMENT

The promotion of sustainable development nowadays permeates the most significant transformations in economic, social, and institutional systems.

It has become increasingly evident that the capitalist market economy, if allowed to operate without adequate constraints, tends to generate distortions that undermine planetary stability and social cohesion. Unregulated economic growth exacerbates inequalities, depletes natural capital and the regenerative capacity of ecosystems, and produces negative externalities that, in an interconnected world, spread rapidly. The COVID-19 pandemic dramatically demonstrated how phenomena originating in a single territory can evolve into global risks. It is therefore necessary to rely on legal instruments capable of governing complexity and coordinating public policies toward a socioeconomic model that differs from those currently in place.

The UN 2030 Agenda, with its 17 sustainable development goals, serves as a reference framework for guiding the evolution of societies toward a form of growth that combines social inclusion, economic prosperity, and ecosystem protection. The current development model is no longer viable, neither environmentally nor socially.

In this context, law, in all its branches, assumes a decisive role. By regulating the behavior of States, public administrations, businesses, and citizens, law can steer society toward objectives that the market alone is unable to pursue. Only law can define limits, incentives, responsibilities, and procedures capable of achieving sustainability goals.

And only law—through its balancing techniques and related institutional mechanisms—can resolve the inevitable conflicts that arise among the objectives of sustainable development. Consider, for example, how increasing food production, necessary to reduce hunger and poverty, may conflict with climate mitigation goals or with responsible production and consumption. Equally significant are the conflicts among the interests of different States, often linked to varying levels of economic development and divergent political priorities.

DOI: 10.1201/9781003595700-7

Legal sciences, however, can no longer be conceived as a self-sufficient system; they must engage in dialogue with the economic, ecological, and natural sciences. These disciplines alone can provide the data and models needed to design norms that effectively promote sustainable development. Legal analysis must therefore integrate knowledge tools that make it possible to measure the effects of public policies and private activities on ecosystems and on human well-being.

6.1.1 Sustainable development, ecosystem services, and natural capital: their connection from the perspective of public policy evaluation and related budgetary decisions

Sustainable development depends heavily on the externalities generated by natural capital and ecosystem services, concepts introduced by the ecological sciences that today form an integral part of legal and institutional debate.

Ecosystem services represent the benefits that natural capital provides to humans, including food, water, and raw materials; the regulation of climate; biogeochemical cycles and natural processes; cultural, recreational, and spiritual functions; and supporting processes such as soil formation and photosynthesis.

In recent years, various supranational institutions have begun to investigate and valorize the role of natural capital and ecosystem services in promoting sustainable development. In particular, the aim was to stimulate the integration, in the impact analyses of public policies, of assessments regarding the effects of the latter not only on production (measured by the Gross Domestic Product [GDP]) but also on the health of ecosystems. The application of knowledge tools drawn from ecology and other natural sciences, as well as from economic analysis, would be decisive in this regard.

To cite a few examples of initiatives of this kind, the Millennium Ecosystem Assessment—a report produced in 2005 under the auspices of the United Nations after 4 year of work—constituted the first attempt to systematically describe the state of global ecosystems, taking into account their importance for human well-being.

The report proposed several possible forms of intervention to maintain and improve ecosystem health, limiting the pressures arising from human activity. These interventions concern collective public action, but also the conduct of businesses and individuals.

The report also highlighted the need to integrate ecosystem-impact assessments into all public and private decision-making processes, ensuring that each decision is based on a sound technical foundation. Ecological assessment would not bind political choice, but it would require avoiding irreversible alterations and would oblige decision-makers to justify any temporary impairments in light of explicitly stated general interests. It would then fall to legal systems to identify the instruments capable of ensuring such balancing.

Following the Millennium Ecosystem Assessment, the German government and the European Union, within the framework of the international Convention on Biological Diversity, promoted the establishment of The Economics of Ecosystems and Biodiversity. The project consists of a research program on the connections between biodiversity and ecosystem services. At the same time, it seeks to develop conceptual models to quantify the value—including economic value—of ecosystem services and the losses resulting from their degradation, thereby making visible the costs and benefits that markets tend to ignore. Economic valuation may rely on various criteria, such as willingness to pay, market prices, avoided costs, restoration costs, and production functions.

Also noteworthy is the experience of Wealth Accounting and the Valuation of Ecosystem Services (WAVES), an international cooperation initiative led by the World Bank aimed at promoting consideration of natural capital in States' development plans and public accounts. The initiative seeks, above all, to develop methodologies for Natural Capital Accounting (NCA) based on shared standards. The objective is to integrate the index normally used to assess national wealth—GDP—with an index that reflects changes in the value of a country's natural resources, that is its natural capital.

The European Union has likewise adopted a distinctive approach in this field, as demonstrated, in particular, by the Seventh Environment Action Program and the Biodiversity Strategy to 2020. These initiatives constitute the reference framework for EU environmental policies and their objectives. To achieve these objectives, the use of methodologies such as environmental accounting and the incorporation—within public decision-making processes—of considerations regarding ecosystem health and related green infrastructure measures is envisaged.

The analytical approach and the economic valuation of environmental issues therefore hold a particularly significant role. They offer a unified and broadly comprehensive vision of the relationship between humankind and nature. In this way, it becomes possible to measure the capacity to generate ecosystem services—now and in the future—while limiting the potentially irreversible negative impacts of human activity.

Avoiding unrealistic proposals for a return to subsistence economies typical of pre-modern eras—which could no longer meet the vital needs of today's global population—these methodologies make it possible to identify solutions that steer economic development while protecting ecosystems and promoting their restoration, where necessary.

From an EU perspective, the introduction of instruments such as NCA—most recently through Regulation (EU) 2024a/3024 of the European Parliament and of the Council of 27 November 2024, amending Regulation (EU) No 691/2011 to introduce new environmental-economic accounting modules—undoubtedly marks a step in this direction. The Commission already employs environmental-economic accounting tools for *ex ante* and *ex post* impact assessments relating to environmental, agricultural, and climate policies.

However, EU budgetary policies still leave no room for assessments that take into account the effects of public policies on natural capital, in addition to traditional macroeconomic indicators (especially GDP, a parameter that implicitly presupposes a context—now disproven by scientific evidence—of unlimited or impact-neutral resources).

Similarly, the EU procedure for monitoring—and in some cases, partially co-deciding—national budgets is based exclusively on indicators referring to GDP, without considering impacts on natural capital or on indices of equitable and sustainable well-being. Indeed, the core objective of EU public-finance policy is defined by the need to achieve a specific ratio between public-budget deficit and public debt, on the one hand, and GDP, on the other.

Until the need to include these dimensions in the system for evaluating public accounts and budgetary policies is recognized at the European level, the instruments currently available —however meritorious—risk producing only limited effects in terms of real transformative capacity.

6.2 SUSTAINABLE DEVELOPMENT AND GREEN DEAL: CLIMATE CHANGE MITIGATION AS A PRIORITY AND NATURE-BASED SOLUTIONS

The expression "sustainable development" emerged in international law, in 1987, and was formally defined in the report "Our Common Future," drafted by the World Commission on Environment and Development under a mandate from the United Nations General Assembly.

The expression evokes the need to rethink the production and consumption models typical of industrial societies, in order to reduce pressure on the planet's natural resources. According to the definition contained in the report, "sustainable development is development that meets the needs of the present without compromising the ability of future generations to meet their own needs."

As it is clear, what is defined is the notion of sustainable development, not the "principle" of sustainable development. Indeed, sustainable development has been affirmed at the international level not so much as a legal principle with specific prescriptive and binding force for all actors, but rather as an objective entrusted to the policies of national governments.

The concept of sustainability entered the EU legal order with the 1992 Maastricht Treaty, which, in Article 2, identified "sustainable growth" respectful of the environment as a model to be promoted by what was then the European Community.

An explicit reference to sustainable development was later introduced in the seventh recital of the preamble to the Treaty on European Union (TEU) by the 1997 Amsterdam Treaty. Sustainable development is considered a principle that Member States must take into account in the context of completing the internal market, strengthening economic and social cohesion, and protecting the environment.

Today, the main references to sustainable development—beyond the preamble of the TEU—are found in Article 3(3) TEU, Article 11 Treaty on the Functioning of the European Union (TFEU), and Article 37 of the Charter of Fundamental Rights of the European Union.

Under the first two provisions, sustainable development appears as a broad political objective, to which environmental protection—characterized by specific public policies—is functionally linked. Achieving this objective requires not only assessing the environmental implications of public policies, but also ensuring a high level of protection and improving environmental quality.

Alongside environmental protection, sustainable development presupposes non-inflationary economic growth; growth that should occur within a model of highly competitive social market economy oriented toward full employment and social progress.

Environmental, economic, and social aspects must therefore be balanced in pursuing sustainable development, but there appears to be an absolute limit: every public policy must ultimately contribute to improve environmental conditions.

Article 37 of the Nice Charter likewise provides that a high level of environmental protection and the improvement of environmental quality must be integrated into Union policies and ensured in accordance with the principle of sustainable development. Yet this provision, too, does not define a precise prescriptive content for the principle.

Thus, even in the EU legal order, sustainable development is essentially a macro-objective entrusted to policymakers—who must define the norms and strategies to give it concrete effect—rather than a prescriptive parameter directly applicable by courts (when assessing the legality of such norms) or by administrations (when implementing environmental policies).

Within the EU legal order, environmental protection—one of the most important expressions of sustainable development—constitutes a transversal policy. Environmental protection implies not only sector-specific rules (on energy, transport, industry, agriculture), but also a reorganization of the criteria for legitimizing public choices. In other words, environmental protection becomes a structural component of the broader strategy for growth and competitiveness, serving as a genuine parameter for policy design and as a condition for accessing a significant share of EU funding.

Climate change mitigation and the ecological transition have become priorities of public action. Both culminate in the European Green Deal and in the construction of a long-term framework oriented toward climate neutrality.

Through the European Green Deal, the European Union seeks to profoundly reshape its development model, steering it toward climate neutrality and environmental sustainability. The Green Deal is built on regulatory measures, investment programs, and governance tools designed to guide the ecological transition of the European economy by mid-century. Its aim is to combine competitiveness, social well-being, and ecosystem protection, outlining a growth trajectory that does not compromise the planet's regenerative capacity.

The rationale of the European Green Deal rests on the need to address the progressive deterioration of climatic and environmental conditions (including rising temperatures, biodiversity loss, and increasing pressure on natural resources). Although this requires substantial investment, the economic and social costs of remedying the damage caused by the degradation of natural habitats and by resource use exceeding ecosystems' regenerative capacity would be far greater.

The Green Deal pursues multiple objectives: achieving climate neutrality by 2050; transitioning to a circular economy; transforming industrial systems toward sustainability and energy efficiency; restoring habitats and eliminating pollution; promoting environmentally respectful agriculture capable of combining food quality and affordability; and ensuring that the transition process as a whole is fair and inclusive.

A wide range of interventions is directed toward achieving these objectives, covering sectors from energy to transport, from pollution to agriculture, and extending to nature and biodiversity restoration.

These latter priorities, in particular, require the implementation of so-called nature-based solutions, defined by the European Commission as

> Solutions that are inspired and supported by nature, which are cost-effective, simultaneously provide environmental, social and economic benefits and help build resilience. Such solutions bring more, and more diverse, nature and natural features and processes into cities, landscapes and seascapes, through locally adapted, resource-efficient and systemic interventions.[1]

There are various types of nature-based solutions, depending on the ecosystem to which they are applied. In urban areas, for example, it is possible to address challenges linked to densification, air pollution, and the effects of climate change through urban parks, ecological corridors, green roofs and walls, sustainable urban drainage systems, and the renaturalization of watercourses. Ecological restoration, reforestation, agroforestry, and sustainable forest management constitute some of the main instruments which, in terrestrial and forest ecosystems, contribute to climate mitigation, hydrogeological stability, and biodiversity conservation.

The restoration of mangroves, coral reefs, and coastal wetlands enables coastal and marine ecosystems to protect shorelines from erosion, reduce the impact of storm surges, and support marine biodiversity. Finally, with regard to agricultural and freshwater ecosystems, nature-based solutions include regenerative agriculture, integrated watershed management, river renaturalization, and the creation of vegetated buffer zones, interventions that improve water quality, reduce soil erosion, increase biodiversity, and strengthen the resilience of rural communities to climate change.

Nature-based solutions are ductile and capable of generating multiple types of benefits, involving a wide range of actors, from public institutions to local communities and the private sector.

It is therefore understandable that the European Union, within the framework of its Green Deal, identifies nature-based solutions as a true cornerstone for the implementation of which it has adopted a dedicated EU Biodiversity Strategy for 2030.[2]

The Strategy was presented by the Commission in May 2020 and is based on the premise that natural capital is rapidly declining and that mere protection is no longer sufficient to reverse this trend. The Strategy seeks to respond, above all, to two critical issues that have emerged over the past 40 year: the decline of approximately 60% in global wildlife populations and the unsatisfactory conservation status—within the Union—of most protected habitats and species.

The Strategy therefore identified several objectives, including the protection of at least 30% of the Union's land and sea areas; strict protection for at least one-third of protected areas, including all primary and old-growth forests; the restoration of degraded ecosystems through a new European legislative framework; a 50% reduction in pesticide use; a 25% increase in organic agricultural land; and the restoration of at least 25,000 km of free-flowing rivers.

The EU's regulatory framework on ecosystem and biodiversity restoration is rather complex.[3] In this regard, a primary role is played by Regulation (EU) 2024/1991 of the European Parliament and of the Council of 24 June 2024 on nature restoration and amending Regulation (EU) 2022/869.

For the first time, the European Nature Restoration Regulation introduces legally binding targets not only to protect what remains of Europe's ecosystems, but also to actively restore what has been degraded, following an organic and proactive logic. The Regulation, in a broad and cross-sectoral manner, requires the restoration of a wide variety of ecosystems. In addition to protected areas and the most vulnerable habitats, it also covers agricultural ecosystems, forests, rivers, urban areas, as well as marine environments and pollinator populations.

The objective of the Regulation—as set out in the provisions of Chapter II—is to restore at least 20% of the Union's land and sea areas by 2030, 60% of habitats by 2040, and 90% of habitats, covering all ecosystems requiring intervention, by 2050.

To achieve the targets laid down in the Regulation—pursuant to the provisions of Chapter III—Member States are required to prepare dedicated national restoration plans. These are strategic planning instruments, adopted consistently with other EU environmental and energy policy tools, and subject to a rigorous assessment process by the European Commission. National restoration plans allow the objectives set by the Regulation to be tailored to the ecological and socioeconomic specificities of each country, while ensuring adequate monitoring of the progress achieved over time.

Each plan must be based on a detailed mapping of degraded ecosystems, enabling the identification of priority areas for intervention. On the basis of this assessment, the plan must define the restoration measures, indicating timelines, operational modalities, responsible actors, and an estimate of the necessary financial coverage.

The results of monitoring activities—based on indicators that are verifiable and comparable at the European level—will allow, in 2033, a review both of the progress made by each Member State in implementing its plan and of the broader effects deriving from the application of the Regulation.

6.3 THE GENERAL PRINCIPLES OF EUROPEAN LAW: PRECAUTION; PREVENTION; RECTIFICATION AT SOURCE OF ENVIRONMENTAL DAMAGE AS A PRIORITY; POLLUTER PAYS; DO NO SIGNIFICANT HARM; INTEGRATION; AND NON-REGRESSION

In EU law, environmental protection is the subject of several general principles.

Article 191(2) TFEU identifies four principles: the precautionary principle, the principle that preventive action should be taken, the principle that environmental damage should be rectified at source as a priority, and the polluter pays principle.

As clarified by the Court of Justice, the precautionary principle requires that risks to public health, safety, and the environment must be prevented, giving priority to the protection of these interests over considerations linked to economic interests.[4]

According to the European Commission, the use of the precautionary principle requires an "identification of potentially negative effects resulting from a phenomenon, product or procedure" and "a scientific evaluation of the risk which because of the insufficiency of the data, their inconclusive or imprecise nature, makes it impossible to determine with sufficient certainty the risk in question."[5]

Against this background, the authorities may choose whether or not to act. In the former case, risk-prevention measures must be proportional to the chosen level of protection, non-discriminatory in their application, consistent with similar measures already taken, based on an examination of the potential benefits and costs of action or lack of action (including, where appropriate and feasible, an economic cost/benefit analysis), subject to review, in the light of new scientific data, and capable of assigning responsibility for producing the scientific evidence.[6]

The prevention principle, instead, requires neutralizing a risk of damage that is certain or highly probable. The risk of damage to be prevented must be scientifically substantiated; in other words, as clarified by the Economic and Social Committee in 2000, the risk must be "measurable or controllable."[7] Risk-mitigation measures must be adopted in accordance with the so-called "best available techniques." The Court of Justice has classified the prevention principle among the fundamental principles of environmental protection.[8] These principles—as set out in the already mentioned Article 191(2) TFEU—shape the action of the Union's institutions, without prejudice to the discretion inherent in the decision-making processes through which the measures for achieving environmental policy objectives are defined[9]; prevention, together with the other fundamental principles, also serves as a criterion used by the Court of Justice when interpreting EU environmental legislation.[10]

The prevention principle constitutes, more or less explicitly, the guiding principle of numerous legislative measures or, in any case, of specific sectors of the Union policy. For example, one could mention Directive 2010/75/EU of the European Parliament and of the Council of 24 November 2010 on industrial emissions (integrated pollution prevention and control) or Directive 2008/98/EC of the European Parliament and of the Council of 19 November 2008 on waste and repealing certain Directives.

The precautionary and prevention principles share the rationale of avoiding the occurrence of environmental harm. The aim is to address environmental issues through an approach which—recognizing that many forms of environmental damage may be irreversible—requires anticipatory action.

However, as noted, the two principles pursue their common objective in different ways. Unlike the prevention principle, the precautionary principle allows the adoption of measures even in the absence of full scientific certainty that a harm will occur. Indeed, during the time needed to obtain scientifically certain or reliable information, irreversible environmental damage could, as mentioned, occur.

The principle according to which environmental damage should be rectified at source as a priority requires that damage or pollution be managed at the place where it occurs. In this way, the aim is to avoid a "relocation elsewhere" of the causes of environmental harm, in order to circumvent controls and the activities intended to prevent or restore environmental damage. Waste management policies, for example, are designed—under principles of self-sufficiency and proximity—against cross-border movements of hazardous waste destined for disposal. As clarified by the Court of Justice, the principle therefore constitutes a specific expression of the principles of self-sufficiency and proximity, which are affirmed at the international level.[11]

At the same time, the principle also has a preventive dimension—Union case-law has emphasized the connection with the prevention principle[12]—serving as a guide for policies aimed at avoiding or reducing, *ex ante*, the sources of environmental harm. This is a preventive aim which, however, in addition to not needing to be linked to an evident risk, is pursued by emphasizing proximity to the source of the harm rather than the integrity of the environment in general terms.[13]

The polluter pays principle entails assigning to the party responsible for pollution the corresponding environmental and social costs. Otherwise, pollution—and the related harm to human health and property—would be absorbed by air and water resources, with the result that society, rather than the polluter, would bear the associated costs.

By contrast, under this principle, the polluter is required to internalize the costs of the pollution generated within the production process. This creates an incentive to favor, in the conduct of one's activities, solutions aimed at environmental protection rather than environmentally harmful behavior.

The polluter pays principle characterizes several legislative measures of the European Union. One may think, for instance, of Directive 2004/35/CE of the European Parliament and of the Council of 21 April 2004 on environmental liability with regard to the prevention and remedying of environmental damage, where the principle is recalled in recital 2 as an instrument for preventing and remedying environmental damage; the operator who has caused environmental damage, or the threat of such damage, must be considered financially responsible, with the aim of encouraging the implementation of measures aimed at minimizing the risks of environmental harm.

The Court of Justice has provided certain interpretative indications regarding the polluter pays principle. In particular, the Court has clarified the need to establish the existence of a causal link between the operator's activity and the pollution produced,[14] as well as that the burdens imposed on the operator responsible for the pollution must be proportionate to the damage caused.[15]

To the principles described thus far, three additional ones may be added, even though they are not mentioned in the Treaties.

First, there is the do no significant harm (DNSH) principle, which is considered for the first time by the Union legislator in the context of sustainable finance. As emerges from Recital 17 of the Regulation (EU) 2019/2088 of the European Parliament and of the Council of 27 November 2019 on sustainability-related disclosures in the financial services sector, the definition of "sustainable investments," from a transparency perspective, must ensure that the undertakings benefiting from such investments "follow good governance practices and the precautionary principle of 'do no significant harm' is ensured, so that neither the environmental nor the social objective is significantly harmed."

Subsequently, the DNSH principle was referred to, from a different perspective, by Regulation (EU) 2020/852 of the European Parliament and of the Council of 18 June 2020 on the establishment of a framework to facilitate sustainable investment and amending Regulation (EU) 2019/2088. This

Regulation, in establishing a regulatory framework aimed at promoting sustainable investments, lays down the conditions that an economic activity must meet in order to benefit from investment and financial support measures of the European Union. To this end, an economic activity must demonstrate, precisely, that it does not cause significant harm to certain environmental objectives identified in Article 17 of the Regulation, namely: climate change mitigation; climate change adaptation; the sustainable use and protection of water and marine resources; the circular economy; pollution prevention and reduction; and the protection and restoration of biodiversity and ecosystems. The DNSH principle was, in turn, recalled by Regulation (EU) 2021/241 establishing the Recovery and Resilience Facility (RRF), whose Article 5 expressly provides that the facility finances only measures that comply with the aforementioned principle. Directly linked to the latter instrument is the related technical guidance for its implementation, prepared by the Commission.[16]

A second principle to be mentioned is the integration principle, which, under Article 11 TFEU, requires that environmental protection requirements must be integrated into the definition and implementation of the Union's policies and actions, particularly with a view to promoting sustainable development.

This principle is likewise recalled in Article 37 of the Charter of Fundamental Rights of the European Union, which provides for the integration of a high level of environmental protection and the improvement of environmental quality into Union policies, as well as their guarantee in accordance with the principle of sustainable development.

The European Commission has encouraged the application of the integration principle in several of its pronouncements. This is the case, for example, of the Communication from the Commission to the Council, the European Parliament, and the European Economic and Social Committee—Integration of Environmental Aspects into European Standardization,[17] which aimed to promote the integration of environmental considerations into Union legislation, in order to fully exploit its potential for harmonizing the laws of the Member States.

The integration principle is therefore primarily intended to guide the exercise of legislative power by the competent European institutions, constituting a source of binding obligations for them. This does not mean that environmental protection must always and necessarily prevail over other interests. Rather, integration requires the Union legislator to assess, on a case-by-case basis and through a balancing exercise, the solutions best suited to achieve the highest possible level of protection within the broader framework of EU policies.[18] To ensure that this occurs, the most important tools to be employed are the *ex ante* assessments and, in particular, environmental impact assessment.

Lastly, a brief reference must be made to the non-regression principle, which requires that the level of environmental protection already achieved not be reduced by subsequent legislative, administrative, or political measures. Although not explicitly mentioned, the Treaties nonetheless provide important indications of the relevance of environmental non-regression. This is particularly the case for the already cited Article 191 TFEU and Article 11 TFEU which—by requiring, respectively, a high level of protection and a continuous improvement in environmental quality, as well as the consideration of environmental protection as a cross-cutting objective of EU policies—implicitly evoke the idea of environmental non-regression. The objectives of the Green Deal and the criteria underlying the DNSH principle, in turn, give operational substance to the aim of preventing any regression in the level of environmental protection achieved thus far.

6.4 THE CHALLENGE OF FINANCIAL SUPPORT FOR GREEN DEAL POLICIES: NEXT GENERATION EU AND OTHER ORDINARY INSTRUMENTS

The investments envisaged by the European Green Deal in innovation, clean technologies, and green infrastructure require the mobilization of a substantial amount of financial resources, for the provision of which the European Union has implemented or established dedicated instruments.

In this regard, the Next Generation EU program stands out: it does not merely supplement an extraordinary public-expenditure plan, but embodies a genuinely alternative vision of collective development, oriented toward sustainability, resilience, and the structural transformation of the Member States' economies.

Next Generation EU (NGEU)is grounded in Decision (EU, Euratom) 2020/2053,[19] by which the Commission was authorized to borrow up to EUR 750 billion on the financial markets, with a view to finance the program's investments.

From a financial perspective, most of the resources of NGEU consist of loans to be repaid, while only a smaller share takes the form of grants. The choice underlying this structure reflects an intergenerational logic: the increase in public investment should foster an economic recovery capable of enabling future generations to repay the debt in a more prosperous and stable context. The program is thus grounded in the idea that sustained growth and the modernization of economic structures constitute the best guarantee for the sustainability of the common debt.

NGEU is organized around six major priorities, ranging from the green transition to digital transformation, from sustainable and inclusive growth to territorial cohesion, and from the strengthening of institutional resilience to policies for younger generations. These pillars should not, however, be considered in isolation; rather, they represent interconnected components of a project that, in general terms, seeks to combine competitiveness, sustainability, and social inclusion.

The concept of sustainability underpinning NGEU must be interpreted in a multidimensional sense, encompassing environmental, social, and economic components. In other words, initiatives financed under NGEU must ensure not only environmental protection and the regeneration of natural resources, but also dignified living conditions, a reduction of inequalities, the fight against poverty, and the promotion of peaceful societies that respect human rights.

To achieve this, national public authorities play a central role. They must regulate markets in a way that favor virtuous economic initiatives, guide operators toward sustainable objectives, and protect the most vulnerable segments of the population. Indeed, the ecological and digital transitions will inevitably entail social costs, such as job losses or difficulties for certain enterprises in adapting to new requirements. Governments therefore have the task of managing, in an orderly and equitable manner, the overall transition process that NGEU seeks to support financially.

Next Generation EU is aimed, above all, at financing the RRF, established by Regulation (EU) 2021/241 of the European Parliament and of the Council of 12 February 2021.

RRF constitutes an innovative instrument, both in terms of its financial scale and its legal-institutional structure, introducing a governance model that combines conditionality, multiannual programming, and a results-oriented approach.

The regulatory framework underpinning the RRF identifies the concept of resilience as its main interpretative key. In this regard, Article 2 of the Regulation defines resilience as "the ability to face economic, social and environmental shocks or persistent structural changes in a fair, sustainable and inclusive way." This definition stems from the awareness that institutions performing well in times of stability may struggle or even collapse in times of crisis. Hence, the need to strengthen not only administrative effectiveness but also the capacity to absorb future shocks, in a context in which further health, climate, and digital crises are anticipated. To achieve this, investment in strong social systems—supported by equally effective healthcare and education systems—is essential.

Access to RRF funds is conditional upon the preparation, by Member States, of the so-called National Recovery and Resilience Plans (NRRPs). These plans are not discretionary instruments; rather, they must comply with strict requirements.

Member States must structure their plans into *milestones* (qualitative results) and *targets* (quantitative results). It is worth noting that 37% of the planned investments must be directed toward climate-neutrality objectives, while 20% must support the digital transition, all in line with the logic underlying the DNSH principle.

The objectives to be pursued through the implementation of the recovery and resilience plans may include reforms aimed at establishing new regulatory frameworks for public services and markets. They may also encompass investment programs for public and private projects designed, for example, to restore existing infrastructure or to build new ones. An important guiding principle, in any case, is to make the Union more resilient and less dependent on current supply chains.

The national plans are subject to assessment by the Commission, which, in accordance with the parameters laid down in Regulation (EU) 2021/241, verifies their content. The Commission then prepares the acts necessary for the implementation of the plans, including a proposal for a Council implementing decision, a Staff Working Document, and the required operational documentation (such as operational and financial agreements).

Following the Commission's proposal, the Council normally has 4 week to adopt the implementing decision. Only after such adoption may the Commission sign the financial agreements with the Member State and proceed with a pre-financing payment of 13%, to be disbursed, where possible, within 2 month. The first pre-financing payments were made in July 2021.

Subsequent disbursements are made once the reforms and investments set out in each national plan have been implemented. Payments are therefore linked to performance, as assessed by the Commission upon request by the Member State. Where the Commission finds that the plan has not been fully implemented, it may proceed with a partial payment, suspending a specific share; in such a case, the Member State may continue implementing the remaining part of its plan. If, however, certain targets or milestones become unattainable due to objective circumstances, the Member State may propose a modification of its plan to the Commission.

Finally, it should be noted that, in exceptional circumstances, any Member State may request the President of the European Council to report on the situation if it considers that another Member State is not properly implementing its national plan.

As anticipated, in addition to establishing a new financing instrument—the RRF—the EU legislator, using NGEU funds, has also strengthened certain pre-existing programs in a targeted manner.

In this regard, with Regulation (EU) 2020/2221,[20] REACT-EU was established as an instrument of recovery assistance, integrated into cohesion policy. It provided for the injection of additional resources into the structural funds, enabling Member States to finance urgent measures related to the pandemic crisis with a level of flexibility significantly higher than usual, including the possibility of 100% EU co-financing.

In the field of research and innovation, NGEU made it possible to increase the budget of Horizon Europe, with a view to supporting projects of high strategic value, particularly in the areas of health, the green transition, and digitalization.

InvestEU was also strengthened, as a program that uses Union guarantees to mobilize private investment. The additional resources from NGEU are managed by the European Investment Bank as well as by national promotional investment institutions. The objective is to expand intervention capacity in four priority areas: sustainable infrastructure, research and digitalization, small and medium-sized enterprises (SMEs), and social investment.

On the climate-transition front, NGEU increases the resources of the Just Transition Fund. The additional funds are allocated to territories most exposed to the effects of decarbonization and are conditional upon the adoption of specific territorial plans. The overall approach remains that of cohesion policy, but with a stronger emphasis on industrial reconversion and worker reskilling.

With regard to RescEU—the European reserve of civil-protection capacities—the aim was to equip the Union with strategic stockpiles and shared operational capabilities to address health and other emergencies. NGEU therefore finances the expansion of reserves and logistical capacities, strengthening a mechanism already provided for under the Civil Protection Mechanism Decision.

Finally, a temporary increase in resources for rural development under the Common Agricultural Policy (CAP) was introduced. The additional funds are directed toward supporting environmental investments, the digitalization of agricultural enterprises, and measures for young farmers, all within the regulatory framework of the European Agricultural Fund for Rural Development (EAFRD).

6.5 THE WASTE MANAGEMENT REGULATION IN THE PERSPECTIVE OF THE CIRCULAR ECONOMY

The circular economy constitutes an economic system model centered on the productive reuse of raw materials and on the minimization of waste generation, with a view to ensuring a more efficient use of the planet's scarce and exhaustible natural resources.

The European Union has made the promotion of a circular economy a true cornerstone of its policies, embarking on the development of a wide range of initiatives in this field.

In 2015, a first action plan on the circular economy was adopted,[21] consisting of 54 actions covering various sectors, including chemicals, plastics, waste, and recycling. Among the main results achieved in the implementation of the plan, one may mention, for example, the entry into force of an updated regulatory framework on waste.[22]

In 2020, a new Circular Economy Action Plan was presented,[23] which was part of the broader European Green Deal strategy. This new plan also outlined several initiatives to promote a circular economy. Consider, for example, the extension of ecodesign to all products,[24] the revision of packaging regulations,[25] and the strengthening of waste shipment rules.[26] Within the framework of the new action plan, additional legislative initiatives were also introduced, completing the structure outlined in 2020; one such example—among others—is the regulation of the right to repair.[27]

It should finally be noted that the developments described thus far have laid the groundwork for the future adoption, in 2026, of a Circular Economy Act. In essence, the measure will establish a Single Market for secondary raw materials, increase the supply of high-quality recycled materials, and stimulate demand for such materials within the EU. The underlying objective is ambitious: to make the Union a global leader in the sector by 2030.

In the context of a waste management inspired by the circular economy model, two aspects appear particularly deserving of attention and further analysis: the introduction and regulation of the so-called extended producer responsibility (EPR) and the notion of end of waste.

The notion of EPR encompasses the obligations and duties imposed on the producer to ensure that they take responsibility for the end-of-life of their products and for the prevention of waste generation, in application of the "polluter pays" principle.[28] The internalization of costs required by EPR should encourage forms of production that are more environmentally sustainable, as these would be comparatively less expensive and therefore allow products to be placed on the market at a lower price.

At the normative level, the first measure to have formally introduced EPR as an instrument aimed at improving waste management and preventing waste generation is Directive 2008/98/EC,[29] which was amended and supplemented by Directive (EU) 2018c/851. The latter intervened to regulate the establishment of extended producer-responsibility schemes, which were to be introduced through measures adopted by the Member States that assign to producers either financial responsibility or both financial and operational responsibility for managing the stage of the life cycle in which the product becomes waste, including separate collection, sorting, and treatment operations (see recital 14).

As of today, therefore, in light of the amendments introduced in 2018, Article 8 of Directive 2008/98/EC allows Member States to adopt measures subjecting those who develop, manufacture, process, treat, sell, or import products to EPR. Such measures may range from the acceptance of returned products and of waste remaining after the product's use to the subsequent management of that waste and the financial responsibility for these activities.

The informational dimension is also taken into account, insofar as it serves to inform the public about the reusability and recyclability of the product.

Article 8 further provides that Member States may adopt measures promoting product design aimed at reducing environmental impact and waste generation, as well as ensuring that waste recovery and disposal take place in accordance with the Directive's requirements.

Directive (EU) 2018c/851 also added Article 8 *bis* to Directive 2008/98/EC, identifying the general minimum requirements for EPR schemes, to apply whenever the measures under Article 8 include the establishment of genuine extended producer-responsibility regimes.

According to the Court of Justice, such schemes

> Are intended to provide a financial and, where appropriate, organizational framework for the management of waste from certain categories of products and thus constitute an essential part of the efficient management of that waste. Indeed, such schemes contribute directly to the attainment of the objective of Directive 2008/98, which is, inter alia, to minimise the negative effects of waste management on the environment and human health.[30]

Among the requirements laid down in Article 8 *bis* for EPR schemes, one may mention, by way of example, the obligation for Member States to clearly define the roles and responsibilities of the actors to whom EPR may apply. These include producers, organizations implementing extended-producer-responsibility obligations on their behalf, public and private waste management operators, local authorities, reuse or preparation-for-reuse operators, and social-economy enterprises.

EPR may permeate all regulatory frameworks concerning waste management. This is the case, for example, of Directive 2012/19/EU,[31] concerning waste electrical and electronic equipment (WEEE). Article 13 of that Directive requires Member States to ensure that the responsibility for financing the costs relating to the management of waste arising from photovoltaic panels is borne by producers for panels placed on the market after 13 August 2012.

A second issue to be examined in relation to waste management inspired by the promotion of a circular economy is that of *end of waste*, an expression referring to the process through which a substance or object ceases to be classified as waste and is consequently reintroduced into the production and commercial cycle.

Here too, the main normative reference is Directive 2008/98/EC (as amended and supplemented by Directive (EU) 2018c/851). In particular, Article 6 of that Directive requires Member States to ensure that materials which have undergone recycling or recovery may cease to be considered "waste."

The provision identifies several conditions that must be met for this to occur, such as compliance with technical requirements for specific purposes or the absence of overall negative impacts on the environment or human health.

As the identification of criteria for obtaining end of waste status is entrusted to the Member States, the Commission plays a monitoring and coordinating role with respect to the initiatives adopted by national competent authorities and may resort to specific implementing acts.

It is also noteworthy that, in the absence of specific measures adopted by either the Member States or the Commission to regulate certain types of situations, national authorities may decide on a case-by-case basis, assessing whether, in the circumstances, the conditions for acquiring end of waste status are met.[32]

As with what has been described for EPR, the discipline of end of waste provides the framework within which to situate the measures adopted to define the criteria for acquiring such status in relation to specific categories of waste. This task has been carried out, for instance, by Regulation (EU) No 333/2011,[33] with regard to metal scrap, by Regulation (EU) No 1179/2012,[34] with regard to glass cullet, and by Regulation (EU) No 715/2013,[35] with regard to copper scrap.

6.6 THE REGULATION OF THE PRODUCTION AND CONSUMPTION OF ENERGY FROM RENEWABLE SOURCES

The reduction of CO_2 emissions constitutes a fundamental objective pursued at both the international and EU levels, in view of ensuring the environmental sustainability of development. With the signing of the 2015 Paris Agreement, the European Union committed itself to contributing to keeping the increase in global temperature well below 2°C, while striving to limit it to 1.5°C.

Within the framework of its Green Deal, the European Union has set particularly ambitious targets, namely a 55% reduction in greenhouse-gas emissions compared to 1990 levels, as well as the achievement of climate neutrality by 2050. These objectives have been made legally binding on the Member States by Regulation (EU) 2021/1119, which establishes the European Climate Law.[36]

A central role in achieving this overarching objective is played by the so-called energy transition, namely the process of replacing energy derived from the use of fossil fuels with "clean" energy, that is, energy produced from renewable energy sources (RES). An economic system centered on the production and consumption of energy generated from renewable sources is explicitly regarded by the European Union as instrumental to realizing sustainable development, particularly with regard to its environmental dimension.

The regulatory framework in this field is rather complex and is centered, first and foremost, on Directive (EU) 2018e/2001 (the so-called RED II),[37] which establishes the general framework for the promotion of RES, setting for 2030 a target of a 32% share of gross final energy consumption from renewables for the Union, to be achieved by each Member State.

RED II has also intervened to promote the development of advanced biofuels, as well as district heating and cooling systems based on the use of renewable energy. The role of Renewable Energy Communities (RECs) has likewise been recognized[38] as new actors within the energy system; specific incentive schemes have been established in their favor.

RED II has been revised by Directive (EU) 2023/2413 (the so-called RED III).[39] The instrument, in particular, has raised the binding 2030 target to 42.5%, while also redefining the sector-specific objectives for industry, buildings, heating/cooling, and transport.

A more detailed framework has likewise been set out with regard to bioenergy.

Moreover, particular attention has been devoted to the establishment of mechanisms for accelerating authorization procedures and to the identification of so-called "acceleration areas," namely areas particularly suitable for the installation of renewable energy facilities.

Several of the interventions introduced by RED III stem from the political impetus generated by the adoption, by the Commission in 2022, of the REPowerEU plan.[40] The plan, aimed at reducing the Union's dependence on Russian fossil fuels, has led to a reshaping of the NRRPs. In other words, part of the Next Generation EU funds—particularly those deriving from permanent loans—has been redirected toward objectives such as energy savings, the diversification of energy supplies, and the rapid deployment of renewable energy. To achieve this, an amendment of the governance framework laid down in Regulation (EU) 2018/1999 became necessary, and this was carried out precisely through RED III.

In addition to having stimulated the adoption of RED III, the REPowerEU plan has also led to the adoption of Council Regulation (EU) 2022/2577,[41] in force until 30 June 2025, with which an acceleration of the deployment of renewable energy has been pursued; in this regard, pursuant to Article 1 of the instrument, the planning, construction, and operation of renewable energy production plants, their grid connection, the grid itself, and storage facilities have been considered to be of overriding public interest and of interest for public health and safety in the balancing of legal interests, particularly, for example, in assessments relating to the conservation of natural habitats.

The development of RES receives further impetus from the measures governing energy markets. In this regard, particular reference must be made to Regulation (EU) 2019/943,[42] which promotes a model of the internal electricity market designed to facilitate the use of renewable energy in various respects, including the priority access of such energy to the grid; Directive (EU) 2019/944,[43] instead, has focused on the role of Citizen Energy Communities, understood as entities capable of producing, consuming, storing, and selling electricity, as well as providing aggregation and demand-management services. The directive thus requires Member States to remove regulatory and administrative barriers that hinder the participation of citizens and communities in electricity markets, ensuring them non-discriminatory access to networks and markets. Moreover, the directive governs both individual and collective self-consumption, recognizing consumers' right to install renewable energy systems, share the energy produced, and participate in collective self-consumption schemes within buildings or neighborhoods.

Finally, further attention must be devoted to the aforementioned Regulation (EU) 2018/1999 on the Union's energy governance, which—aiming to make electricity-market reforms an integral part of coherent national policies and to ensure the effective integration of renewables—has required Member States to draw up Integrated National Energy and Climate Plans. These plans must include specific measures for the promotion of RES, the flexibility of the electricity system, and the development of energy communities.

NOTES

1 European Commission, European Research Executive Agency, "Nature-based solutions. EU-funded NbS research projects tackle the climate and biodiversity crises."
2 Communication from the Commission to the European Parliament, the Council, the European Economic and Social Committee, and the Committee of the Regions EU Biodiversity Strategy for 2030: bringing nature back into our lives (COM(2020) 380).
3 One may think to Directive 2009/147/EC of the European Parliament and of the Council of 30 November 2009 on the conservation of wild birds, to Council Directive 92/43/EEC of 21 May 1992 on the conservation of natural habitats and of wild fauna and flora, to Directive 2000/60/EC of the European Parliament and of the Council of 23 October 2000 establishing a framework for Community action in the field of water policy, and to Directive 2008/56/EC of the European Parliament and of the Council of 17 June 2008 establishing a framework for community action in the field of marine environmental policy (Marine Strategy Framework Directive).
4 Court of First Instance, Second Chamber, Extended Composition, 26 November 2002, Joined Cases T-74/00, T-76/00, T-83/00 to T-85/00, T-132/00, T-137/00, and T-141/00, Artegodan GmbH e aa. v Commission of the European Communities.
5 Communication from the Commission on the precautionary principle (COM(2000) 1 final).
6 *Ivi.*
7 Opinion of the Economic and Social Committee on the "*Use of the precautionary principle*" (2000/C/268/04), point 2.7.
8 CJEU, 9 December 2008, C-121/07, Commission of the European Communities v French Republic.
9 See, for example, CJEU, 14 July 1998, C-284/95, Safety Hi-Tech Srl v S. & T. Srl.
10 See, for example, CJEU, 2 October 2003, C-322/00, Commission of the European Communities v Kingdom of the Netherlands.
11 CJEU, 9 July 1992, C-2/90, Commission of the European Communities v Kingdom of Belgium.
12 CJEU, 9 December 2008, C-121/07, Commission v France.
13 According to the Court of Justice, for example, the obligation on Member States to adopt measures to reduce sulphur dioxide and nitrogen emissions must not be interpreted in light of the general environmental situation of the area in which the industrial installation concerned is located (CJEU, 7 July 2005, C-364/03, Commission of the European Communities v Hellenic Republic).

14 CJEU, 9 March 2010, C-378/08, Raffinerie Mediterranee (ERG) SpA and others v Ministero dello Sviluppo economico and others.
15 CJEU, 16 July 2009, C-254/08, Futura Immobiliare srl Hotel Futura, and others v Comune di Casoria.
16 Commission Notice. Technical guidance on the application of "do no significant harm" under the Recovery and Resilience Facility Regulation (C/2023/111).
17 COM/2004/0130.
18 CJEU, 14 July 1998, C-284/95, Safety Hi-Tech Srl v S. & T. Srl.
19 Council Decision (EU, Euratom) 2020/2053 of 14 December 2020 on the system of own resources of the European Union and repealing Decision 2014/335/EU, Euratom.
20 Regulation (EU) 2020/2221 of the European Parliament and of the Council of 23 December 2020 amending Regulation (EU) No 1303/2013 as regards additional resources and implementing arrangements to provide assistance for fostering crisis repair in the context of the COVID-19 pandemic and its social consequences and for preparing a green, digital, and resilient recovery of the economy (REACT-EU).
21 Communication from the Commission to the European Parliament, the Council, the European Economic and Social Committee, and the Committee of the Regions. Closing the loop—An EU action plan for the Circular Economy (COM(2015) 614 final).
22 In 2018, a "waste package" was adopted, consisting of four directives through which an equal number of legislative acts were amended. This refers to Directive (EU) 2018a/849 of the European Parliament and of the Council of 30 May 2018, amending Directives 2000/53/EC on end-of-life vehicles, 2006/66/EC on batteries and accumulators and waste batteries and accumulators, and 2012/19/EU on waste electrical and electronic equipment; to Directive (EU) 2018b/850 of the European Parliament and of the Council of 30 May 2018, amending Directive 1999/31/EC on the landfill of waste; to Directive (EU) 2018c/851 of the European Parliament and of the Council of 30 May 2018, amending Directive 2008/98/EC on waste; and to Directive (EU) 2018d/852 of the European Parliament and of the Council of 30 May 2018, amending Directive 94/62/EC on packaging and packaging waste.
23 Communication from the Commission to the European Parliament, the Council, the European Economic and Social Committee, and the Committee of the Regions. A new Circular Economy Action Plan. For a cleaner and more competitive Europe (COM(2020) 98 final).
24 Regulation (EU) 2024a/1781 of the European Parliament and of the Council of 13 June 2024 establishing a framework for the setting of ecodesign requirements for sustainable products, amending Directive (EU) 2020/1828 and Regulation (EU) 2023/1542 and repealing Directive 2009/125/EC.
25 Regulation (EU) 2025/40 of the European Parliament and of the Council of 19 December 2024 on packaging and packaging waste, amending Regulation (EU) 2019/1020 and Directive (EU) 2019/904, and repealing Directive 94/62/EC.
26 Regulation (EU) 2024b/1157 of the European Parliament and of the Council of 11 April 2024 on shipments of waste, amending Regulations (EU) No 1257/2013 and (EU) 2020/1056 and repealing Regulation (EC) No 1013/2006.

27 Directive (EU) 2024/1799 of the European Parliament and of the Council of 13 June 2024 on common rules promoting the repair of goods and amending Regulation (EU) 2017/2394 and Directives (EU) 2019/771 and (EU) 2020/1828.
28 This principle is enshrined in Article 191(2) TFEU.
29 Directive 2008/98/EC of the European Parliament and of the Council of 19 November 2008 on waste and repealing certain Directives.
30 CJEU, 10 July 2025, C-254/23, INTERZERO, and others.
31 Directive 2012/19/EU of the European Parliament and of the Council of 4 July 2012 on waste electrical and electronic equipment (WEEE).
32 A similar situation arose in the case decided by the Court of Justice on 24 October 2019 (C-212/18), Prato Nevoso Termo Energy Srl, where it was clarified that national authorities must, in any event, establish criteria in a generally applicable national legal act. In addition, a Member State may, taking into account all the relevant factors and the most recent scientific and technical knowledge, decide not to establish, as regards certain types of waste, either criteria or the possibility of an individual decision recognising end of waste status.
33 Council Regulation (EU) No 333/2011 of 31 March 2011 establishing criteria determining when certain types of scrap metal cease to be waste under Directive 2008/98/EC of the European Parliament and of the Council.
34 Commission Regulation (EU) No 1179/2012 of 10 December 2012 establishing criteria determining when glass cullet ceases to be waste under Directive 2008/98/EC of the European Parliament and of the Council.
35 Commission Regulation (EU) No 715/2013 of 25 July 2013 establishing criteria determining when copper scrap ceases to be waste under Directive 2008/98/EC of the European Parliament and of the Council.
36 Regulation (EU) 2021/1119 of the European Parliament and of the Council of 30 June 2021 establishing the framework for achieving climate neutrality and amending Regulations (EC) No 401/2009 and (EU) 2018/1999 ("European Climate Law").
37 Directive (EU) 2018e/2001 of the European Parliament and of the Council of 11 December 2018 on the promotion of the use of energy from renewable sources.
38 As is well known, pursuant to Article 2 of Directive 2018/2001, Renewable Energy Communities constitute autonomous associative entities in which natural or legal persons, including public authorities, may participate. Their objective is to benefit the members of the relevant community along one of the three dimensions of sustainability—environmental, economic, or social—through the sharing of self-generated energy produced by renewable energy installations. Moreover, renewable energy communities, in addition to consuming within the community the renewable energy they produce, may also sell (and, under national legislation, are required to transfer) the share of energy that is not self-consumed, it being understood that their primary objective cannot be the distribution of financial profits.
39 Directive (EU) 2023/2413 of the European Parliament and of the Council of 18 October 2023 amending Directive (EU) 2018e/2001, Regulation (EU) 2018/1999 and Directive 98/70/EC as regards the promotion of energy from renewable sources, and repealing Council Directive (EU) 2015/652.
40 COM(2022) 230.
41 Council Regulation (EU) 2022/2577 of 22 December 2022 laying down a framework to accelerate the deployment of renewable energy.

42 Regulation (EU) 2019/943 of the European Parliament and of the Council of 5 June 2019 on the internal market for electricity.
43 Directive (EU) 2019/944 of the European Parliament and of the Council of 5 June 2019 on common rules for the internal market for electricity and amending Directive 2012/27/EU.

BIBLIOGRAPHY

CJEU, 9 July 1992, C-2/90, Commission of the European Communities v Kingdom of Belgium.

CJEU, 14 July 1998, C-284/95, Safety Hi-Tech Srl v S. & T. Srl.

CJEU, 2 October 2003, C-322/00, Commission of the European Communities v Kingdom of the Netherlands.

CJEU, 7 July 2005, C-364/03, Commission of the European Communities v Hellenic Republic.

CJEU, 9 December 2008a, C-121/07, Commission of the European Communities v French Republic.

CJEU, 9 December 2008b, C-121/07, Commission v France

CJEU, 16 July 2009, C-254/08, Futura Immobiliare srl Hotel Futura and others v Comune di Casoria.

CJEU, 9 March 2010, C-378/08, Raffinerie Mediterranee (ERG) SpA and others v Ministero dello Sviluppo economico and others.

CJEU, 24 October 2019 (C-212/18), Prato Nevoso Termo Energy Srl.

CJEU, 10 July 2025, C-254/23, INTERZERO and others.

Commission Notice. Technical guidance on the application of 'do no significant harm' under the Recovery and Resilience Facility Regulation (C/2023/111).

Commission Regulation (EU) No 1179/2012 of 10 December 2012 establishing criteria determining when glass cullet ceases to be waste under Directive 2008/98/EC of the European Parliament and of the Council.

Commission Regulation (EU) No 715/2013 of 25 July 2013 establishing criteria determining when copper scrap ceases to be waste under Directive 2008/98/EC of the European Parliament and of the Council.

Communication from the Commission on the precautionary principle (COM(2000) 1 final).

Communication from the Commission to the Council, the European Parliament and the European Economic and Social Committee - Integration of Environmental Aspects into European Standardisation (COM/2004/0130).

Communication from the Commission to the European Parliament, the Council, the European Economic and Social Committee and the Committee of the Regions. Closing the loop - An EU action plan for the Circular Economy (COM(2015) 614 final).

Communication from the Commission to the European Parliament, the Council, the European Economic and Social Committee and the Committee of the Regions EU Biodiversity Strategy for 2030. Bringing nature back into our lives (COM(2020a) 380).

Communication from the Commission to the European Parliament, the Council, the European Economic and Social Committee and the Committee of the Regions. A

new Circular Economy Action Plan. For a cleaner and more competitive Europe (COM(2020b) 98 final).
Communication from the Commission to the European Parliament, the European Council, the Council, the European Economic and Social Committee and the Committee of the Regions REPowerEU Plan (COM(2022) 230).
Council Decision (EU, Euratom) 2020/2053 of 14 December 2020 on the system of own resources of the European Union and repealing Decision 2014/335/EU, Euratom.
Council Regulation (EU) No 333/2011 of 31 March 2011 establishing criteria determining when certain types of scrap metal cease to be waste under Directive 2008/98/EC of the European Parliament and of the Council.
Council Regulation (EU) 2022/2577 of 22 December 2022 laying down a framework to accelerate the deployment of renewable energy.
Court of First Instance, Second Chamber, Extended Composition, 26 November 2002, Joined Cases T-74/00, T-76/00, T-83/00 to T-85/00, T-132/00, T-137/00 and T-141/00, Artegodan GmbH e aa. v Commission of the European Communities.
Directive 2004/35/EC of the European Parliament and of the Council of 21 April 2004 on environmental liability with regard to the prevention and remedying of environmental damage.
Directive 2008/98/EC of the European Parliament and of the Council of 19 November 2008 on waste and repealing certain Directives.
Directive 2009/147/EC of the European Parliament and of the Council of 30 November 2009 on the conservation of wild birds, to Council Directive 92/43/EEC of 21 May 1992 on the conservation of natural habitats and of wild fauna and flora, to Directive 2000/60/EC of the European Parliament and of the Council of 23 October 2000 establishing a framework for Community action in the field of water policy, to Directive 2008/56/EC of the European Parliament and of the Council of 17 June 2008 establishing a framework for community action in the field of marine environmental policy (Marine Strategy Framework Directive).
Directive 2010/75/EU of the European Parliament and of the Council of 24 November 2010 on industrial emissions (integrated pollution prevention and control) (recast).
Directive 2012/19/EU of the European Parliament and of the Council of 4 July 2012 on waste electrical and electronic equipment (WEEE).
Directive (EU) 2018a/849 of the European Parliament and of the Council of 30 May 2018 amending Directives 2000/53/EC on end-of-life vehicles, 2006/66/EC on batteries and accumulators and waste batteries and accumulators, and 2012/19/EU on waste electrical and electronic equipment.
Directive (EU) 2018b/850 of the European Parliament and of the Council of 30 May 2018 amending Directive 1999/31/EC on the landfill of waste.
Directive (EU) 2018c/851 of the European Parliament and of the Council of 30 May 2018 amending Directive 2008/98/EC on waste.
Directive (EU) 2018d/852 of the European Parliament and of the Council of 30 May 2018 amending Directive 94/62/EC on packaging and packaging waste.
Directive (EU) 2018/2001 of the European Parliament and of the Council of 11 December 2018e, on the promotion of the use of energy from renewable sources.
Directive (EU) 2019/944 of the European Parliament and of the Council of 5 June 2019, on common rules for the internal market for electricity and amending Directive 2012/27/EU.

Directive (EU) 2023/2413 of the European Parliament and of the Council of 18 October 2023 amending Directive (EU) 2018/2001, Regulation (EU) 2018/1999 and Directive 98/70/EC as regards the promotion of energy from renewable sources, and repealing Council Directive (EU) 2015/652.

Directive (EU) 2024/1799 of the European Parliament and of the Council of 13 June 2024 on common rules promoting the repair of goods and amending Regulation (EU) 2017/2394 and Directives (EU) 2019/771 and (EU) 2020/1828.

European Commission, European Research Executive Agency, "Nature-based solutions. EU-funded NbS research projects tackle the climate and biodiversity crises" (n.d.).

Opinion of the Economic and Social Committee on the *"Use of the precautionary principle"* (2000/C/268/04).

Regulation (EU) 2019/943 of the European Parliament and of the Council of 5 June 2019 on the internal market for electricity.

Regulation (EU) 2020/2221 of the European Parliament and of the Council of 23 December 2020 amending Regulation (EU) No 1303/2013 as regards additional resources and implementing arrangements to provide assistance for fostering crisis repair in the context of the COVID-19 pandemic and its social consequences and for preparing a green, digital and resilient recovery of the economy (REACT-EU).

Regulation (EU) 2021/1119 of the European Parliament and of the Council of 30 June 2021 establishing the framework for achieving climate neutrality and amending Regulations (EC) No 401/2009 and (EU) 2018/1999 ('European Climate Law').

Regulation (EU) 2024a/1781 of the European Parliament and of the Council of 13 June 2024 establishing a framework for the setting of ecodesign requirements for sustainable products, amending Directive (EU) 2020/1828 and Regulation (EU) 2023/1542 and repealing Directive 2009/125/EC.

Regulation (EU) 2024b/1157 of the European Parliament and of the Council of 11 April 2024 on shipments of waste, amending Regulations (EU) No 1257/2013 and (EU) 2020/1056 and repealing Regulation (EC) No 1013/2006.

Regulation (EU) 2025/40 of the European Parliament and of the Council of 19 December 2024 on packaging and packaging waste, amending Regulation (EU) 2019/1020 and Directive (EU) 2019/904, and repealing Directive 94/62/EC.

Index

Note: Pages in '*italics*' refer to figures, pages in '***bold-italics***' refer to boxes, pages in '**bold**' refer to tables and pages followed by n refer to notes.

For Product Safety Concerns and Information please contact our EU representative GPSR@taylorandfrancis.com
Taylor & Francis Verlag GmbH, Kaufingerstraße 24, 80331 München, Germany

www.ingramcontent.com/pod-product-compliance
Lightning Source LLC
LaVergne TN
LVHW011448121026
849885LV00032B/296
9781032975740